Power System Analysis

Power System Analysis

S. Ramar

Professor and Head
Department of Electrical and Electronics Engineering
SACS M.A.V.M.M. Engineering College
Madurai
Former Director
Central Electricity Authority

S. Kuruseelan

Assistant Professor
Department of Electrical and Electronics Engineering
Bannari Amman Institute of Technology
Sathyamangalam

PHI Learning Private Limited

Delhi-110092
2013

POWER SYSTEM ANALYSIS
S. Ramar and S. Kuruseelan

ISBN-978-81-203-4733-5

The export rights of this book are vested solely with the publisher.

Published by Asoke K. Ghosh, PHI Learning Private Limited, Rimjhim House, 111, Patparganj Industrial Estate, Delhi-110092 and Printed by Raj Press, New Delhi-110012.

To
Our Beloved Parents

To

Our Beloved Parents

Contents

Foreword

I am very happy to write this foreword for the book *Power System Analysis* by Prof. S. Ramar, Head, Electrical and Electronics Engineering Department, SACS M.A.V.M.M. Engineering College, Madurai and Prof. S. Kuruseelan, Assistant Professor, Electrical and Electronics Engineering Department, Bannari Amman Institute of Technology, Sathyamangalam.

The authors have taken concerted efforts to write this book, covering all the aspects of Power System Analysis comprehensively. The contents of the book are presented in simple, precise and systematic manner with lucid explanation so that the students can easily understand the underlying principles and build up their knowledge brick-by-brick. A number of examples and diagrams have been included to aid conceptual understanding of the subject. This book can be used as a textbook for the Electrical Engineering course at undergraduate level and reference book for postgraduate courses.

Having been myself a practicing engineer and teacher in the Engineering field for more than five decades, I am sure that this book would not only meet the requirements of students but also can serve as reference book to the teachers. The book will be useful to the practicing engineers as well. I feel that the students can preserve this book for future reference and career.

I congratulate the authors for bringing out such a comprehensive book on the subject and wish them success in their endeavour.

Prof. P. Manjunatha Rao
B.E., M.Sc. (Engg.), M.I.E., M.I.S.T.E.

Director
SACS M.A.V.M.M. Engineering College
Madurai

Preface

In the last two decades, significant development has taken place in the field of power systems. Computer control of large interconnected areas for efficient and reliable operation under all conditions of operation has become essential. Power System Analysis has become an important aspect of power system planning, operation and control.

There are many books available on the topic Power System Analysis. The aim of this book is to integrate the largely scattered information in this field and provide a ready reference. This book is mainly based on our lecture notes to undergraduate students and can serve as a textbook to undergraduate students. A number of problems have been solved to clearly illustrate the concepts. The problems given as exercise can be given as assignments for the practice of the students. Each chapter is appended with short answer questions with answers.

Chapter 1 gives an overview of the power system covering the basic components, per phase analysis of balanced three-phase system, per unit values and application. Chapter 2 covers modelling of generator, transformer, transmission line and loads and impedance diagram.

Chapter 3 covers the power flow analysis in detail that includes classification of bus, bus admittance matrix and formation of power flow equations. Chapter 4 deals with methods of solving power flow equations namely Gauss–Seidal method, Newton–Raphson method, Fast Decoupled Power Flow method and comparison of these three methods. The methods are described with the help of algorithm and flow chart for easy understanding of students.

Chapter 5 is confined to balanced fault analysis using circuit theory principles, while Chapter 6 describes systematic balanced fault analysis using bus impedance matrix including algorithm for building bus impedance matrix.

Chapter 7 explains the concept used for analyzing unbalanced or unsymmetrical fault, the symmetrical components, sequence impedances and networks. Chapter 8 describes computation of fault currents due to unsymmetrical faults like line to ground fault, line to line fault and double line to ground fault. The chapter also covers unbalanced fault analysis using bus impedance matrix.

Chapter 9 deals with power system stability. It covers various concepts of stability such as rotor angle and voltage stability, mid term and long term stability and steady state (static and dynamic) and transient stability. The chapter also covers equal area criterion to study transient stability and factors affecting stability. Chapter 10 discusses numerical methods like Euler method, modified Euler method and Runge–Kutta method to solve swing equations. Multi-machine stability analysis has also been covered.

GATE questions and answers pertaining to the topic on Power System Analysis are appended to help the students scale competitive examinations successfully.

Although utmost care has been taken to make the book free of errors both in text as well as in solved examples, yet the authors feels obliged if any errors present are brought to their notice. Constructive criticism of the book is warmly welcome.

S. Ramar
S. Kuruseelan

Acknowledgements

We sincerely thank Prof. P. Manjunatha Rao, Director, SACS M.A.V.M.M. Engineering College, who has been the constant source of inspiration and guidance in all our efforts and for his kind foreword.

Wide reference has been made to a number of books in the power system field and we are indebted to the authors of these books.

We sincerely thank our colleagues and students who have given their ideas and suggestions to bring this book in a user-friendly format.

Last but not the least, we are grateful to PHI Learning, our Publisher and Mr. Mayur Joseph, Assistant Production Manager for printing the book in the present form.

S. Ramar
S. Kuruseelan

CHAPTER 1

Introduction

Being a clean and versatile form of energy and easy to transmit over long distances, electric power is the most preferred form of energy. Electricity is a vital input for economic and social development in our society. Besides its importance in the growth of a country's economy, it plays a major role in the life of a common man and has a direct impact on the quality of life. Its demand has been growing faster than many other forms of energy. The demand for electricity in India has been growing at an average growth rate of 7 to 8% and demand-supply gap has widened over the years. Providing reliable and inexpensive electricity is essential for economic development of the country and better standard of living of the people.

1.1 Power System Structure

The power system consists of three basic components, namely generation, transmission and distribution. Transmission lines connect generating stations, distribution systems and other power systems while the distribution systems connect loads to transmission lines at substations. Substations perform transformation and switching functions by means of transformers and circuit breaker.

Generation, transmission and distribution systems are connected in a power system so as to form a coherent power system structure for optimal operation. The structure of the power system is shown in Figure 1.1.

Power is generated by various types of generating plants such as hydro, thermal (coal, gas, oil-fired), nuclear and wind. The generation voltage is generally in the range of 11 kV to 25 kV. To transmit large quantum of power over long distances with low transmission losses, the generated voltage is stepped up to high/extra high/ultra high voltage (220/400/765 kV) transmission system. Small generators meeting local load may, however, be connected to sub-transmission system. The transmission system not only handles large blocks of power but also inter connects all the generating stations and power systems, forming a grid network. Very large customers may avail power directly from the transmission level. Transfer of energy to other pool members may also take place in this level. The power can be routed in any direction on the various links of the transmission system in a way to achieve overall operating economy.

The sub-transmission circuit distributes power to a number of distribution substations in a certain area at a voltage level that typically varies between 66 kV and 110/132 kV. It receives

1

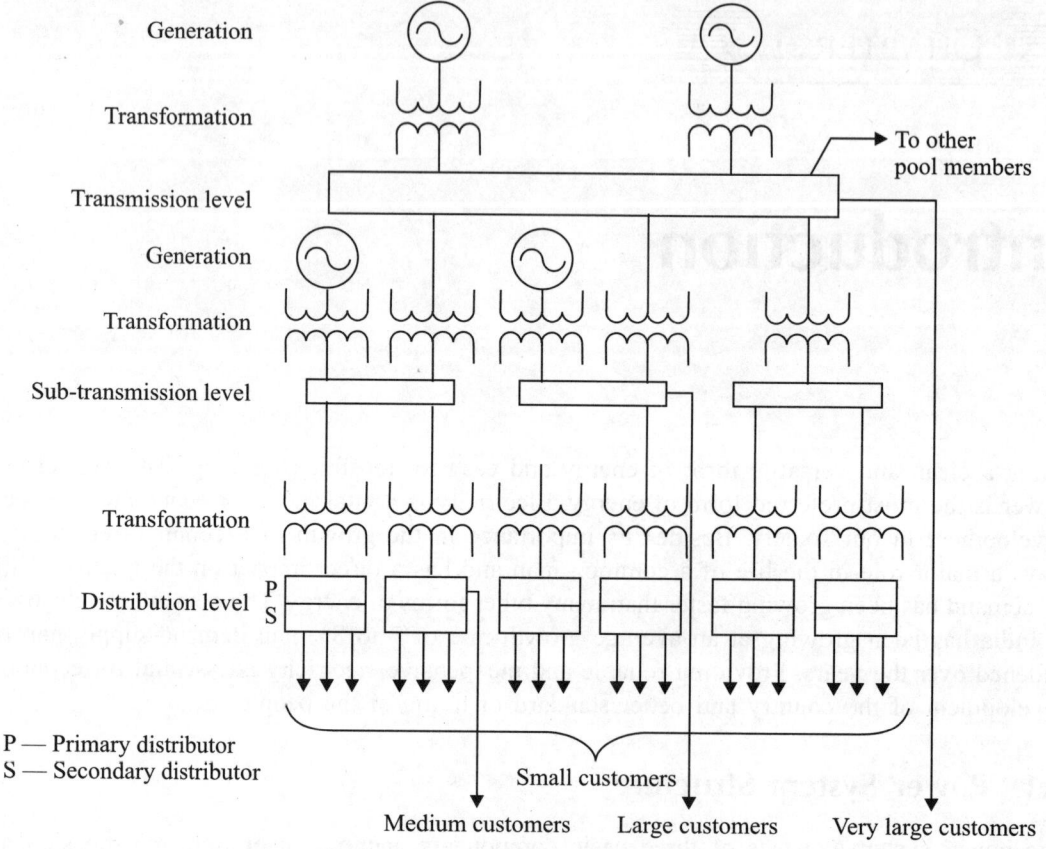

Figure 1.1 Power system structure.

power directly from the generating stations or via bulk power substations. Large customers are served directly from these stations. When system grows and load density increases, transmission level may become sub-transmission level.

The distribution circuits are at the lower most voltage level—primary or feed voltage level (33/11 kV) and secondary or consumer voltage level (230/415 V). They supply power to medium industrial customers at primary level and small (domestic/agricultural/commercial/small industry) customers at secondary level. Distribution engineering covers variety of services like over head line or underground cable, metering, switching and protection.

While the transmission level has a loop structure, the sub-transmission and distribution level is usually in the radial mode of operation.

1.2 Growth of Power System

Pearl Street system in New York established by Edison around 1880 is the first power system. Power was generated at 100 V by steam-turbine-driven dc dynamos and was distributed through underground cables to light around four hundred incandescent bulbs each of 80 W. Invention of transformer and induction motor in 1884–88 gave impetus to develop ac system and made it possible to increase ac voltage for transmission of power and decrease for distribution. Single-phase ac system started in 1884 and three-phase system in 1891. The transmission voltage increased progressively from 15 kV to 765 kV, and 1050 kV. The systems of various countries in Europe started integrating to their advantage. During the early stage, ac systems of different frequencies ranging from 25 to 133 1/3 Hz were in place. While lower frequencies are desirable for transmission to reduce the effect of reactance, higher frequencies are preferable for meeting the load like incandescent lamps. After due consideration, as a compromise, the frequency was standardized to 60 Hz in United States and 50 Hz in Europe and Asia.

In India, a hydro project near Darjeeling was first set-up around 19th century end and a 4.5 MW Sivasamudram hydro project in Karnataka was commissioned in 1902. The transmission voltage steadily increased from 132 kV in 1931, 220 kV in 1962, 400 kV in 1978 and to 765 kV by 20th century end.

High Voltage Direct Current (HVDC) transmission system was first introduced in the world by 1954 through the construction of 100 kV, 20 MW system in Sweden. The highest transmission voltage as on date is +/– 750 kV.

In India, the first HVDC Bi-pole line namely +/– 500 kV, 1500 MW Rihand – Dadri Bi-pole was commissioned in 1987 followed by Vindhyachal back-to-back station (2 × 250 MW) in 1988.

The electric generation in India has increased many fold since independence. The generating capacity at the time of India becoming republic was hardly 1,700 MW. This has increased tremendously to 1,99,877 MW in the year 2012 even without 31,516 MW captive plants owned by non-utilities. The capacity increase over the period is shown in Figure 1.2.

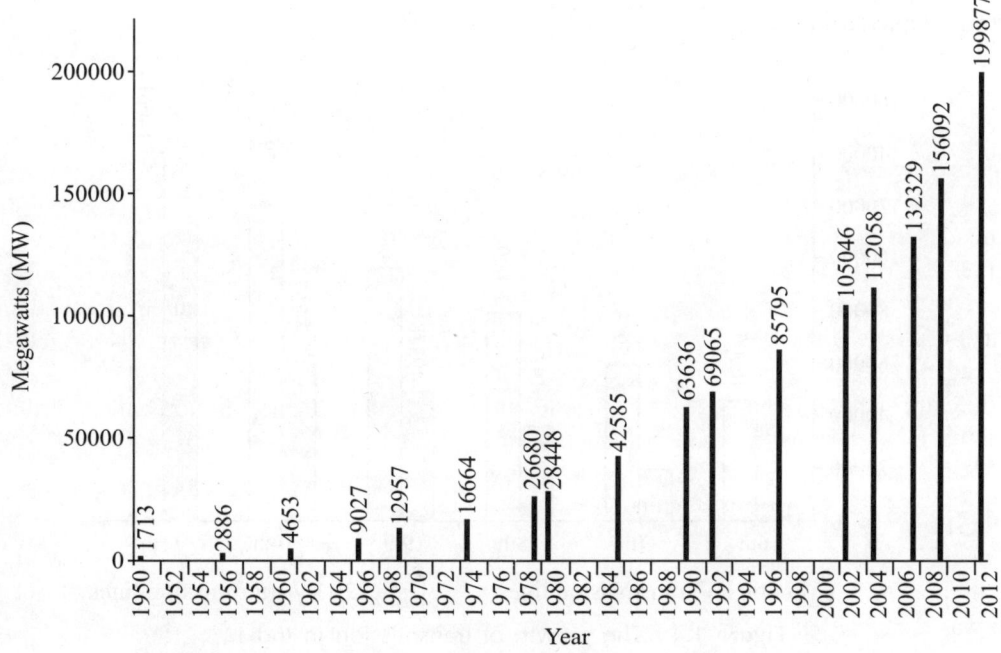

Figure 1.2 Growth of installed generating capacity in India.

The generating capacity of 1,99,877 MW is owned by Utilities comprising 38,990 MW hydro, 1,31,604 MW thermal, 4,780 MW nuclear and 24,503 MW Renewable Energy Sources (RES) and 31,516 MW captive plants is owned by non-utilities as shown in Figure 1.3.

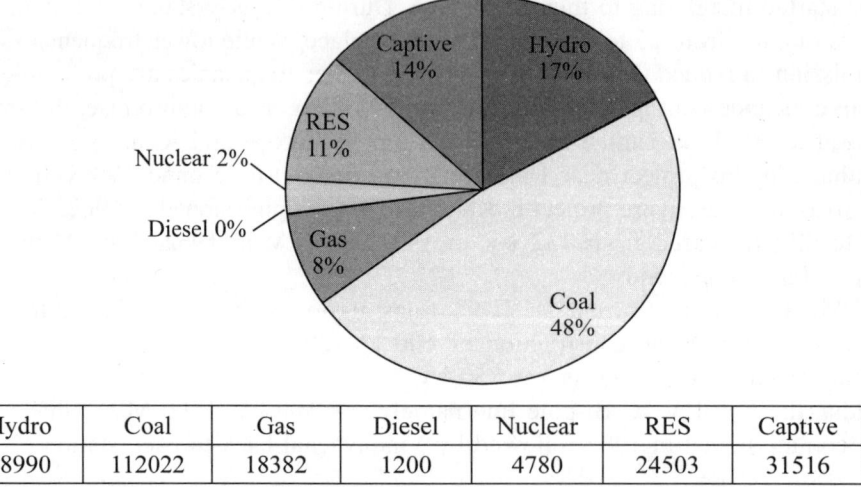

Hydro	Coal	Gas	Diesel	Nuclear	RES	Captive
38990	112022	18382	1200	4780	24503	31516

Figure 1.3 All India installed generating capacity as on 31-03-2012.

The transmission and distribution line length had also corresponding increase. The 400 kV EHV transmission line length increased from 6029 ckm in 6th plan to 1,13,367 ckm by the end of 11th plan and 220 kV transmission line length from 46,005 ckm to 1,40,614 ckm. This growth is depicted in Figure 1.4.

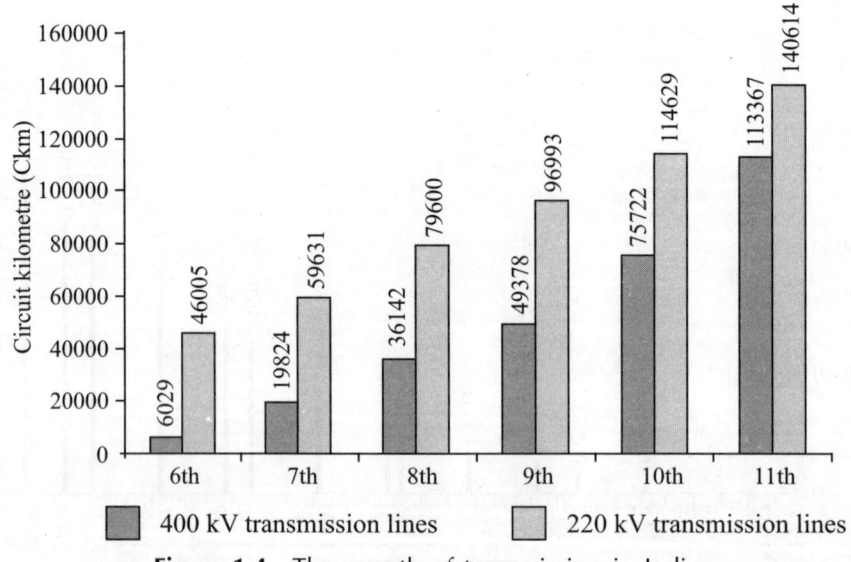

Figure 1.4 The growth of transmission in India.

The power system was developed in the form of clusters in the initial stage with generation and distribution meeting the local loads. The following advantages of interconnecting power systems led to the integrated operation of the various systems.

Advantages of interconnecting power systems

(a) With the interconnection of systems, the diversity factor improves and the installed generating capacity (number of generators) required to supply the combined system load is less.
(b) Since the system becomes large, it is highly stable with better frequency and voltage regulation.
(c) The spinning reserve requirement gets reduced.
(d) Bigger unit size can be planned resulting in the economy of scale.
(e) Better generation mix improves the economies in operation.
(f) Emergency assistance from neighbouring system(s) can be availed.
(g) Optimum location of new plants is possible.

The interconnecting power systems has certain disadvantages also.

Disadvantages of interconnecting power systems

(a) Increase in short-circuit level causes increased breaker capacity and duty.
(b) Any disturbance in one system may spread to other systems.
(c) The tie-line interconnection needs to be provided with proper protection arrangements.
(d) Maintaining synchronism among all generators is difficult.
(e) Proper coordination among power systems is important.

These are, however, surmountable with suitable design and coordination.

1.3 Power System—Planning and Operation

Reliable and uninterrupted service to loads is achieved through proper planning and operation. The following power system tools are used for this purpose:

- Load or Power flow studies
- Short-circuit studies
- Stability studies
- System protection and Relay coordination studies

Power flow studies

Power flow studies are carried out to determine the voltage magnitude and angle at all the buses and active and reactive power flow over all the lines and transformers under steady state. These are the initial values for further studies such as short-circuit and stability studies. These studies are useful to identify unacceptable voltage conditions and overloading, effect of line/transformer outage and alternative plans. The power flow studies cover load forecasting, planning studies and operational studies.

Short-circuit studies

Short-circuit studies are carried out to study the behaviour of the system under fault conditions—various types of faults and at various locations. The main objectives of the studies are:

- To determine current interrupting capacity of circuit breakers.
- To determine relay settings and coordination.
- To design grounding system.
- To calculate current limiting reactors if required.

Stability studies

Stability Studies are conducted to ensure system stability on the occurrence of faults and after their clearance. The types of studies cover rotor angle stability and voltage stability. The stability studies can also be classified as steady state (Static and Dynamic) stability and transient stability.

Under steady state, current and voltage are sinusoidal functions with constant magnitude and frequency, while under transient state current and voltage vary and change from one state to another due to fault or switching.

The instantaneous voltage (v) is given by

$$v = V_m \sin \omega t \tag{1.1}$$

where,

V_m = maximum magnitude of voltage
ω = angular frequency

The current comprises steady state and transient components.

$$i = i_{ss} + i_{tr} \tag{1.2}$$

where,

$$Z = \sqrt{R^2 + X^2} \quad \text{and} \tag{1.3}$$

$$\theta = \tan^{-1} \frac{X}{R} \tag{1.4}$$

$$i_{ss} = \left(\frac{V_m}{Z}\right) \sin(\omega t - \theta) \tag{1.5}$$

$$i_{tr} = \left(\frac{V_m}{Z}\right) \sin\theta\, e^{\frac{-Rt}{L}} \tag{1.6}$$

$$i = \left(\frac{V_m}{Z}\right) [\sin(\omega t - \theta) + \sin\theta\, e^{\frac{-Rt}{L}}] \tag{1.7}$$

Although the impedance during transient state will be less than the steady state value, same impedance has been assumed in the above analysis.

Transient value, in this case the dc off-set current, decays as time passes, but steady state value continues.

The fault current during transient state and steady state conditions is shown in Figure 1.5.

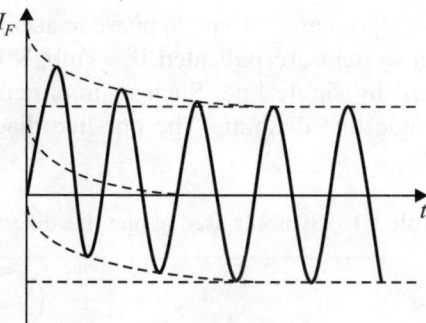

Figure 1.5 Fault current.

1.4 Per Phase Analysis

The generation is by three-phase synchronous generators. The transformation and transmission are by three-phase transformers and three-phase lines. The distribution is through three-phase network. Thus complete phase symmetry is ensured by design and balanced distribution of single-phase loads, as shown in Figure 1.6.

In a balanced system, generator and load neutrals are at same potential. Neutral current is, therefore, zero; neutral impedance has no effect. Thus complete decoupling between the three phases is ensured. In other words, solution of a single-phase network completely determines the solution of three-phase network (Figure 1.7).

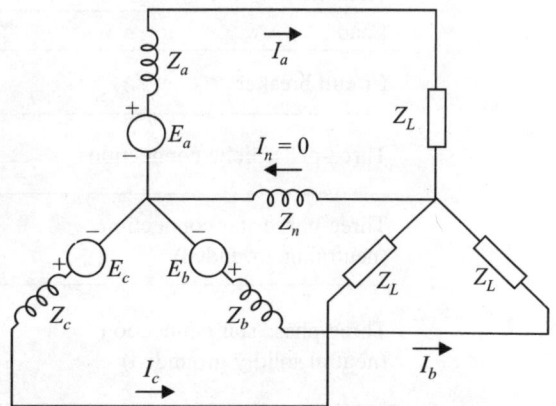

Figure 1.6 Balanced three-phase network.

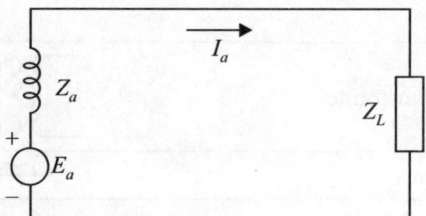

Figure 1.7 Single-phase equivalent of balanced three-phase network of Figure 1.6.

For the reference phase a

$$E_a = (Z_a + Z_L)I_a \tag{1.8}$$

The voltages in the other phases have the same magnitude but are progressively shifted in phase by 120 degrees. The current will also have similar property.

1.5 One-line Diagram

Three-phase balanced system is represented by single phase in accordance with per phase analysis. The components of the power system are indicated by symbols (Table 1.1). For example, the transmission line is represented by single line. Such a simplified diagram of power system is called one-line diagram or single line diagram. The one-line diagram gives information about the system in concise form.

Table 1.1 Symbols used in one-line diagram

Rotating machine	
Two-winding transformer	
Transmission line	
Load	
Circuit breaker	
Three-phase delta connection	
Three-phase star connection (neutral ungrounded)	
Three-phase star connection (neutral solidly grounded)	
Three-phase star connection (neutral grounded through reactor)	
Three-winding transformer	
Current transformer	
Voltage transformer	

The details given on the diagram depend on the type of study. For power flow study, the location of the circuit breakers is not needed. But for short-circuit and stability studies it is required. The ratings with usual nomenclature are:

- Generators: S, V, X
- Transformers: V_p/V_s, S, X
- Load: P, Q or S, pf
- Transmission line: X

A typical one-line diagram indicating the symbols and ratings of power system components is shown in Figure 1.8.

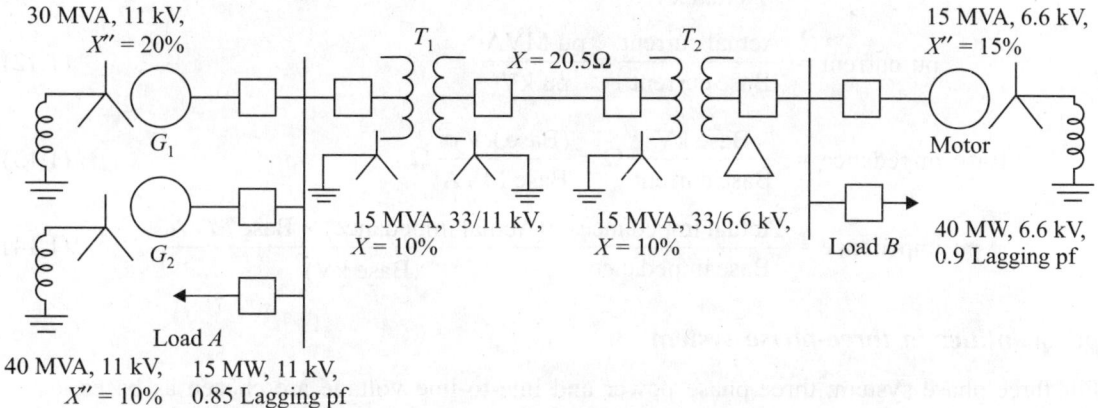

Figure 1.8 A typical one-line diagram.

1.6 Per-unit (pu) Quantities

The solution of power system having different voltage levels and power capacities requires normative (per-unit) values, which are easily amenable for computer calculations. Further, where equipment rating is not known it can be easily assumed because pu values lie in a narrow range.

The basic power system parameters are voltage, current, apparent power and impedance. Of these four, two are independent while the remaining two are dependent and can be computed from the independent quantities. Usually, voltage in kV and apparent power in MVA are selected as base quantities.

The per-unit (pu) quantity is the ratio of actual to base value expressed in decimal. Sometimes the ratio is expressed in per cent, which is 100 times the pu value. The pu value representation is better than per cent value, because multiplication of say two per-unit values gives correct per-unit value of the combination, but the multiplication of two per cent values gives incorrect per cent value of the combination and to get the correct value the product is to be divided by 100.

pu quantities in single phase system

For chosen base voltage (kV) and base apparent power (MVA) in single-phase system

$$\text{pu power} = \frac{\text{Actual MVA}}{\text{Base MVA}} \tag{1.9}$$

$$\text{pu voltage} = \frac{\text{Actual kV}}{\text{Base kV}} \tag{1.10}$$

$$\text{Base current} = \frac{\text{Base MVA}}{\text{Base kV}} \text{ kA} \tag{1.11}$$

$$\text{Actual current} = \frac{\text{Actual MVA}}{\text{Actual kV}} \text{ kA}$$

$$\text{pu current} = \frac{\text{Actual current}}{\text{Base current}} = \frac{\text{pu MVA}}{\text{pu kV}} \tag{1.12}$$

$$\text{Base impedance} = \frac{\text{Base kV}}{\text{Base current}} \, \Omega = \frac{(\text{Base kV})^2}{\text{Base MVA}} \, \Omega \tag{1.13}$$

$$\text{pu impedance} = \frac{\text{Actual impedance}}{\text{Base impedance}} = \frac{(\text{Actual impedance}) \times \text{Base MVA}}{(\text{Base kV})^2} \tag{1.14}$$

pu quantities in three-phase system

For three-phase system, three-phase power and line-to-line voltage are chosen as bases.

$$\text{pu voltage} = \frac{\text{Actual kV}_l}{\text{Base kV}_l} = \frac{\sqrt{3} \times \text{Actual kV}_{ph}}{\sqrt{3} \times \text{Base kV}_{ph}} = \frac{\text{Actual kV}_{ph}}{\text{Base kV}_{ph}} = \frac{\text{Actual kV}}{\text{Base kV}} \tag{1.15}$$

$$\text{pu power} = \frac{\text{Actual MVA}_{3ph}}{\text{Base MVA}_{3ph}} = \frac{3 \times \text{Actual MVA}_{ph}}{3 \times \text{Base MVA}_{ph}} = \frac{\text{Actual MVA}_{ph}}{\text{Base MVA}_{ph}} = \frac{\text{Actual MVA}}{\text{Base MVA}}$$

$$\tag{1.16}$$

$$\text{Base current} = \frac{\text{Base MVA}_{3ph}}{\sqrt{3} \times \text{Base kV}_l} \text{ kA} \tag{1.17}$$

$$\text{Actual current} = \frac{\text{Actual MVA}_{3ph}}{\sqrt{3} \times \text{Actual kV}_l} \text{ kA}$$

$$\text{pu current} = \frac{\text{Actual current}}{\text{Base current}} = \frac{\text{Actual MVA}_{3ph}}{\sqrt{3}(\text{Actual kV}_l)} \times \frac{\sqrt{3}\,(\text{Base kV}_l)}{\text{Base MVA}_{3ph}} = \frac{\text{pu MVA}}{\text{pu kV}} \tag{1.18}$$

Impedance is a per phase quantity and is calculated as follows for a three-phase system:

$$\text{Base current/phase} = \frac{\text{Base MVA}_{3ph}}{\sqrt{3} \times \text{Base kV}_l} \text{ kA}$$

$$\text{Base impedance} = \frac{\text{Base kV}_{\text{ph}}}{\text{Base current}}\,\Omega = \frac{\left(\dfrac{\text{Base kV}_l}{\sqrt{3}}\right) \times \sqrt{3} \times \text{Base kV}_l}{\text{Base MVA}_{3\text{ph}}}$$

$$= \frac{(\text{Base kV}_l)^2}{\text{Base MVA}_{3\text{ph}}} = \frac{(\text{Base kV})^2}{\text{Base MVA}} \tag{1.19}$$

$$\text{pu impedance} = \frac{\text{Actual impedance}}{\text{Base impedance}} = \frac{\text{Actual impedance} \times \text{Base MVA}}{(\text{Base kV})^2} \tag{1.20}$$

Thus, the pu formulas are same irrespective of single phase or three-phase system. Only difference is the formula for base current.

1.7   Changing Base of Per-unit Quantities

The manufacturers of equipment usually give the pu impedance on the basis of its own rating. However, the base quantities of the system may be different. In order to bring the pu impedance of the equipment on the common system base, it has to be changed.

$$Z_{\text{pu}_{\text{given}}} = Z_{\text{actual}} \times \frac{(\text{Base MVA}_{\text{given}})}{(\text{Base kV}_{\text{given}})^2}$$

$$Z_{\text{pu}_{\text{new}}} = Z_{\text{actual}} \times \frac{(\text{Base MVA}_{\text{new}})}{(\text{Base kV}_{\text{new}})^2}$$

By dividing and rearranging, we get

$$Z_{\text{pu}_{\text{new}}} = Z_{\text{pu}_{\text{given}}} \times \frac{(\text{Base kV}_{\text{given}})^2}{(\text{Base kV}_{\text{new}})^2} \times \frac{\text{Base MVA}_{\text{new}}}{\text{Base MVA}_{\text{given}}} \tag{1.21}$$

For transformer pu impedance seen from either side is same. The transformer pu impedance seen from primary side is

$$Z_{\text{pu}_p} = Z_{\text{actual}_p} \times \frac{(\text{Base MVA}_p)}{(\text{Base kV}_p)^2}$$

$$= Z_{\text{actual}} \times \frac{(\text{Base kV}_s)^2}{(\text{Base kV}_p)^2} \times \frac{\text{Base MVA}_p}{(\text{Base kV}_s)^2}$$

(Multiplying numerator and denominator by $(\text{Base kV}_s)^2$)

$$= Z_{\text{actual}_s} \times \frac{(\text{Base MVA}_s)}{(\text{Base kV}_s)^2}$$

(transferring impedance from primary to secondary side and using the property that MVA seen from either side of the transformer is same)

$$= Z_{\text{pu}_s} \text{(transformer pu impedance seen from secondary side)}$$

Advantages of pu system

The pu system has the following advantages:

- Manufacturers usually provide impedance values in pu.
- The pu impedances of machines of same type and widely different ratings lie within a narrow range, whereas ohmic values differ very much. Where equipment rating is not known it can, therefore, be easily assumed.
- Per-unit values referred to either side of transformer remains same. pu impedance of a three-phase transformer is independent of the type of winding connection, star or delta.
- Computational efforts are reduced very much.
- The chance of confusion between line and phase quantities is greatly reduced since the factors of $\sqrt{3}$ and 3 are eliminated.

Disadvantages of pu system

The pu system has the following disadvantages:

- Familiar equations in unscaled circuit get modified when scaled into pu, i.e. $\sqrt{3}$ and 3 are removed.
- Equivalent circuits of components get modified making them more abstract, i.e. Phase-shift present clearly in unscaled circuit vanishes in pu circuit.

Points for solving pu problems

The following points are kept in view while solving problems using pu values:

  (i) Base kilovolt and base megavolt ampere are selected in one part of the system, usually the largest part. (Base values for three-phase system are line-to-line voltage and three-phase megavolt ampere)

 (ii) For other parts of the system (other sides of transformers), the base kilovolt is according to line-to-line voltage ratios of transformers, but the base megavolt ampere remains same in all parts of system.

(iii) Impedance information is usually given in per-unit or percent based on its own rating. The pu impedance given on base other than chosen base is to be changed to base values chosen. If impedance value is given in ohm, pu impedance is calculated from the base values chosen.

(iv) For three single-phase transformers connected as a three-phase unit, the three-phase ratings are determined from single-phase ratings as given below.

Base MVA = MVA_{3ph} = 3 × MVA_{1ph}

Base voltage = $kV_l = \sqrt{3}\,kV_{ph}$ for star connected side = kV_{ph} for delta connected side

The pu impedance of three-phase unit is same as that of the single-phase units.

EXAMPLE 1.1   The primary and secondary windings of a single-phase 1000 kVA, 2000/1000 V transformer have leakage reactances each of 2 Ω. Find the pu reactance of the transformer.

Solution   The per-unit impedance of the transformer can be obtained either first by transferring both primary and secondary impedances to any one side of the transformer and finding the per-unit value or by finding the per-unit value on the respective side and adding them, as the pu value is same seen by either side. Using the second method,

$$Z_{pu} = Z_{actual} \times \frac{\text{Base VA}}{\text{Base V}^2}$$

$$X_p = 2 \times \frac{1000 \times 10^3}{2000^2} = 0.5 \text{ pu}$$

$$X_s = 2 \times \frac{1000 \times 10^3}{1000^2} = 2.0 \text{ pu}$$

$$X_T = X_p + X_s = 0.5 + 2.0 = 2.5 \text{ pu}$$

EXAMPLE 1.2 The one-line diagram of a three-phase system is shown in Figure 1.9. The transformer reactance is 20% on a base of 100 MVA, 23/115 kV and the line impedance is $Z = j66.125 \ \Omega$. The load at bus 2 is $S_2 = 184.8$ MW $+ j6.6$ MVAR and at bus 3 is $S_3 = 0$ MW $+ j20$ MVAR. It is required to hold the voltage at bus 3 at $115 \angle 0°$ kV. Working in per-unit, determine the voltage at buses 2 and 1.

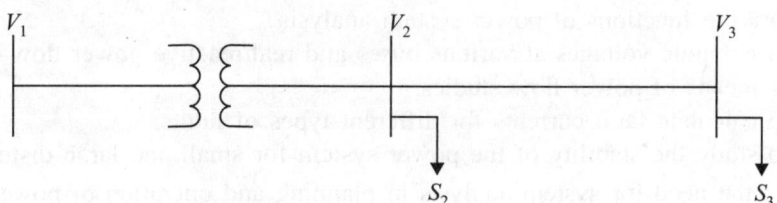

Figure 1.9 Example 1.2.

Solution Let the base values be

$S_b = 100$ MVA and $V_b = 115$ kV on line side

$V_{b3} = V_{b2} = 115$ kV and

$V_{b1} = 115 \times \dfrac{23}{115} = 23$ kV as per transformation

$V_3 = \dfrac{115 \angle 0°}{115} = 1 \angle 0° \text{ pu}$

$S_3 = \dfrac{0 + j20}{100} = j0.2 \text{ pu}$

$S_2 = \dfrac{184.8 + j6.6}{100} = 1.848 + j0.066 \text{ pu}$

$Z_1 = j66.125 \times \dfrac{100}{(115)^2} = j0.5 \text{ pu}$

$Z_T = \dfrac{j20}{100} = j0.2 \text{ pu}$

$$I_3 = \frac{S_3^*}{V_3^*} = \frac{-j0.2}{1} = -j0.2$$

$$V_2 = V_3 + I_3 Z_1 = 1 - j0.2 \times j0.5 = 1 + 0.1 = 1.1\angle 0° \text{ pu}$$
$$= 1.1 \times 115 = 126.5\angle 0° \text{kV}$$

$$I_2 = \frac{S_2^*}{V_2^*} = \frac{1.848 - j0.066}{1.1} = 1.68 - j0.06 \text{ pu}$$

$$I_1 = I_3 + I_2 = -j0.2 + 1.68 - j0.06 = 1.68 - j0.26$$
$$V_1 = I_1 Z_T = 1.1 + (1.68 - j0.06) \times j0.2$$
$$= 1.1 + 0.052 + j0.336 = 1.152 + j0.336 = 1.2\angle 22.42° \text{ pu}$$
$$= 1.2 \times 23 = 27.6\angle 22.42° \text{ pu}$$

Short Answer Questions

1. What is power system analysis?

   The evaluation of power system is called power system analysis.

2. What are the functions of power system analysis?
   (i) To compute voltages at various buses and real/reactive power flow between buses by means of power flow studies.
   (ii) To calculate fault currents for different types of faults.
   (iii) To study the stability of the power system for small and large disturbances.

3. What is the need for system analysis in planning and operation of power system?

   Reliable and uninterrupted service to loads is achieved through proper planning and operation.
   (i) Power system comprising generation, transmission and distribution has to be planned to meet the demand forecast for future years reliably.
   (ii) Generation dispatch has to be scheduled to match with the varying demand and losses optimally and securely.

4. What are the components of power system?

   The components of power system are generators, transformers, transmission lines, substations, distribution systems and loads.

5. Define per-unit value.

   The per-unit value of any quantity is defined as the ratio of actual quantity to its base quantity expressed as decimal. The ratio in per cent is 100 times the value in per-unit.

6. What is the need for base values?

   The components in various sections of power system may operate at different voltage and power levels. It will be convenient for power system analysis if the voltage, power, current and impedance ratings of these components are expressed with reference to a common value called base value.

7. How are the base values chosen in per-unit representation of a power system?

Any two of the following four quantities are chosen as base values and the balance two are computed:

Voltage, current, apparent power and impedance. Usually, voltage and apparent power are selected as base quantities.

Base apparent power in MVA or VA: The rated apparent power of the largest section or part is chosen as base. This base remains same in all sections of the system.

Base voltage in kV or V: The rated voltage of the largest section is chosen as base. The base voltages of remaining sections are assigned depending on the turns ratio of the transformers connecting the sections.

The base values of current and impedance are calculated from these chosen base values.

8. What are the advantages of per-unit representation?

   (i) Manufacturers usually provide impedance values in pu.

   (ii) Per-unit impedances of machines of same type and widely different ratings lie within a narrow range, whereas ohmic values differ very much. Where equipment rating is not known it can, therefore, be easily assumed.

   (iii) Per-unit values referred to either side of transformer remains same. Per-unit impedance of a three-phase transformer is independent of the type of winding connection, star or delta.

   (iv) Computational efforts are reduced very much.

   (v) The chance of confusion between line and phase quantities is greatly reduced since the factors of $\sqrt{3}$ and 3 are eliminated.

9. Write the equation for per-unit impedance.

$$\text{pu impedance} = \frac{\text{Actual impedance}}{\text{Base impedance}}$$

$$= \frac{\text{Actual impedance} \times \text{Base impedance}}{(\text{Base kV})^2}$$

10. Write the equation for per-unit impedance if change of base occurs.

$$Z_{pu_{new}} = Z_{pu_{given}} \times \frac{(\text{Base kV}_{given})^2}{(\text{Base kV}_{new})^2} \times \frac{\text{Base MVA}_{new}}{\text{Base MVA}_{given}}$$

11. What is single- or one-line diagram?

A single or one-line diagram is a diagrammatic representation of power system, in which the components are represented by their symbols and the interconnection between them shown by a single straight line (even though the system is three-phase balanced system). The ratings and the impedances of the components are also marked on the diagram.

12. What is the purpose of using single-line diagram?

The purpose of the single-line diagram is to supply in concise form the significant information about the system. The ratings are marked against the components. Further details given on the diagram depend on the type of study.

13. Give the equation for transforming base kV on low voltage (LV) side to high voltage (HV) side of transformer and vice-versa.

$$\text{Base kV on LV side} = \text{Base kV on LV side} \times \frac{\text{HV voltage rating}}{\text{LV voltage rating}}$$

$$\text{Base kV on LV side} = \text{Base kV on LV side} \times \frac{\text{HV voltage rating}}{\text{LV voltage rating}}$$

14. What is the difference between steady state and transient state?

    Steady state is characterized by slow change in system or load and transient state is characterized by sudden change in system or load.

15. The reactance of a generator designated X'' is given as 0.2 per-unit based on the generator's nameplate rating of 20 kV, 500 MVA. The base for calculations is 22 kV, 100 MVA. Find X'' on the new base.

$$Z_{pu_{new}} = Z_{pu_{given}} \times \frac{(\text{Base kV}_{given})^2}{(\text{Base kV}_{new})^2} \times \frac{\text{Base MVA}_{new}}{\text{Base MVA}_{given}}$$

$$X'' = 0.2 \times \left(\frac{20}{22}\right)^2 \times \left(\frac{100}{500}\right)$$

Exercises

1.1 The one-line diagram of a three-phase power system is shown in Figure E1.1. Impedances are marked in pu on a 100 MVA, 400 kV base. The load at bus 2 is $S_2 = 15.93$ MW $- j33.4$ MVAR and at bus 3 is $S_3 = 77$ MW $+ j14$ MVAR.

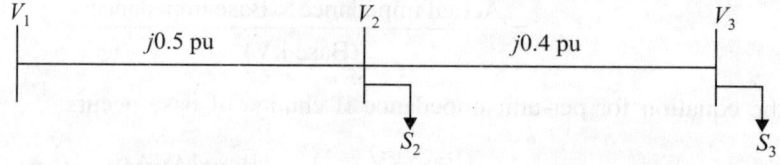

Figure E1.1 One-line diagram for Problem 1.1.

It is required to hold the voltage at bus 3 at 400 $\angle 0°$ kV. Working in per-unit, determine the voltage at buses 2 and 1. [**Ans.** 440 kV, 480 kV]

1.2 Figure E1.2 shows the schematic diagram of a radial transmission system.

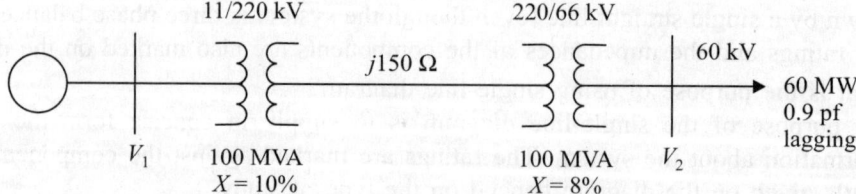

Figure E1.2 Schematic diagram for Problem 1.2.

The ratings and reactances of the various components are shown therein. A load of 60 MW at 0.9 power factor lagging is tapped from the 66 kV substation which is to be maintained at 60 kV. Calculate the terminal voltage of the synchronous machine.

[**Ans.**   12.21 kV]

1.3 A 120 MVA, 19.5 kV generator has a synchronous reactance of 0.15 pu and it is connected to a transmission line through a transformer rated 150 MVA, 230/18 kV (star/delta) with $X = 0.1$ pu.

  (i) Calculate the pu reactance by taking generator rating as base values.

  (ii) Calculate the pu reactance by taking transformer ratings as base values.

  (iii) Calculate the pu reactances for a base value of 100 MVA and 220 kV on HT side of transformer.

[**Ans.**   (i) 0.0681, (ii) 0.22, (iii) 0.1603, 0.0729]

1.4 A simple power system is shown in Figure E1.3. Redraw this system where the per-unit impedances of the components are represented on a common 5000 VA base and common system base value of 250 V.

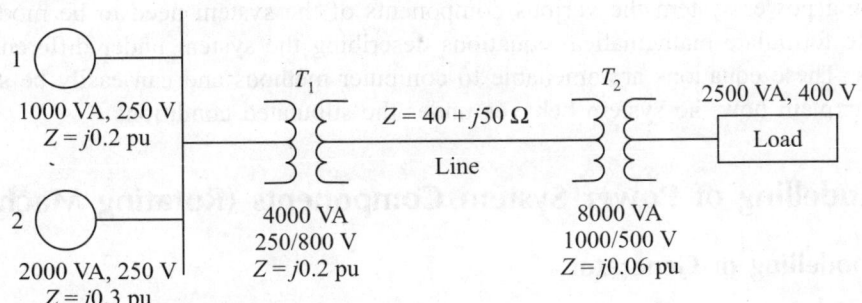

Figure E1.3   One-line diagram for Problem 1.4.

[**Ans.**   1, 0.75, 0.25, 0.313 + j1.17, 0.0585, 0.5 VA pu]

CHAPTER 2

Modelling of Power System Components

For studying power system the various components of the system need to be modelled. This will enable formulate mathematical equations describing the system under different operating conditions. These equations are amenable to computer methods and can easily be solved. The solutions explain how the system behaves under the stipulated conditions.

2.1 Modelling of Power System Components (Rotating Machines)

2.1.1 Modelling of Generator

Under steady state condition, the armature reaction of the synchronous generator reduces magnetic flux in addition to the leakage flux. This effect is represented as a reactance called *armature reaction reactance*, X_a. The sum of leakage reactance, X_l and X_a is called *synchronous reactance*, X_s. The generator voltage in terms of terminal voltage and armature current as shown in Figure 2.1 is given by

$$V = E - IZ_s \quad \text{where } Z_s = R + jX_s$$

Normally, $R \ll X_s$

$$V = E - jIX_s$$

In the case of salient pole generator, saliency effect is usually neglected and

$$X_s = X_d$$
$$V = E - jIX_d$$

In general

$$V = E - jIX \tag{2.1}$$

where

$$X = X_d \text{ (under steady state)}$$
$$= X_d' \text{ (under transient state)}$$
$$= X_d'' \text{ (under sub-transient state)}$$

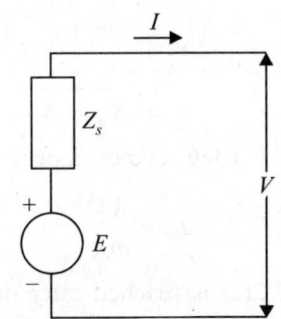

Figure 2.1 Generator model.

At the instant of short-circuit the dc off-set current can induce current in the rotor field winding and in the damper winding causing demagnetizing effect as the current is at zero pf lag. This effect can be represented by two reactances X_f (representating the flux created by induced current in the field winding) and X_{ad} (representating the flux created by induced current in the damper winding) in parallel with X_a, as shown in Figure 2.2(a). The combined effect of all the reactances is to reduce the net reactance of the machine and the short-circuit current is very high in this state, called *sub-transient* state. This reactance is called sub-transient reactance X_d'' and the current sub-transient current I''.

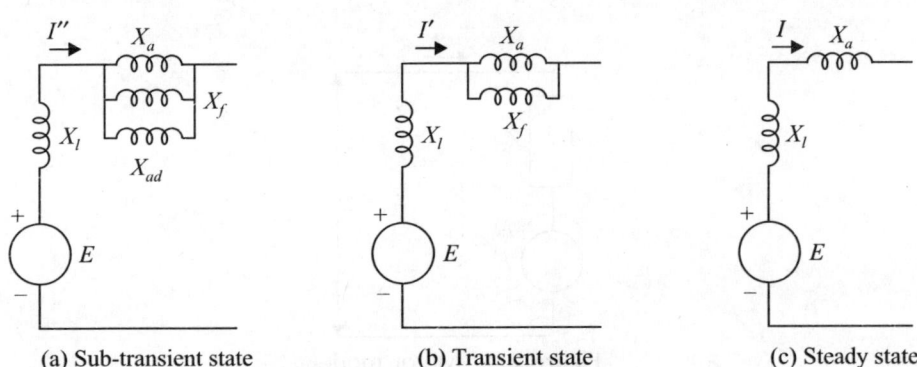

(a) Sub-transient state   (b) Transient state   (c) Steady state

Figure 2.2 Equivalent circuits of synchronous generator.

$$X_d'' = X_l + \cfrac{1}{\cfrac{1}{X_a} + \cfrac{1}{X_f} + \cfrac{1}{X_{ad}}} \tag{2.2}$$

The induced current in the damper winding decays very quickly with a time constant (T_d) given by

$$T_d'' = \frac{X_d''}{\omega R} \tag{2.3}$$

where R is the Thevenin's resistance at the terminals of the network.

After this initial decay, the equivalent circuit is shown in Figure 2.2(b) and this state is called *transient state*, with transient reactance X_d' and transient current I'.

$$X'_d = X_l + \cfrac{1}{\cfrac{1}{X_a} + \cfrac{1}{X_f}} \tag{2.4}$$

The transient state will also exist for a few cycles depending on the time constant given by

$$T'_d = \frac{X'_d}{\omega R} \tag{2.5}$$

The steady state shown in Figure 2.2(c) is reached after that. The reactance is

$$X_d = X_l + X_a \tag{2.6}$$

It may be noted that

$$X''_d < X'_d < X_d \tag{2.7}$$

and

$$I'' > I' > I \tag{2.8}$$

2.1.2 Modelling of Synchronous Motor

The equivalent circuit of synchronous motor is similar to that of generator except that the current flows in the opposite direction as shown in Figure 2.3.

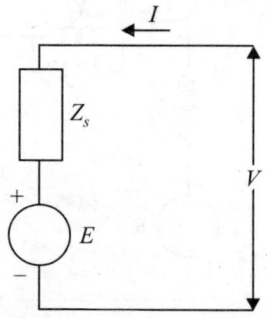

Figure 2.3 Motor model.

$$V = E + IZ_s$$

Normally, $R \ll X_s$

$$V = E + jIX_s$$

By neglecting saliency effect, we get

$$V = E + jIX_d \tag{2.9}$$

On fault, generated emf of synchronous motor load contributes to the fault current and the current flows from motor to fault at voltage V_f.

$$V_f = E - jIX \tag{2.10}$$

where

$X = X_d$ (under steady state)

$\quad = X''_d$ (under sub-transient state)

2.1.3 Modelling of Induction Motor

The equivalent circuit of induction motor is similar to that of synchronous motor

$$V = E + IZ_s$$

Normally, $R << X_s$

$$V = E + jIX_s \tag{2.11}$$

Generated emf of induction motor load contributes to the fault current immediately after the fault during sub-transient conditions only, as there is no field and the current flows from motor to fault at voltage V_f.

$$V_f = E - jIX_d'' \tag{2.12}$$

2.1.4 Modelling of Static Loads

Generation and absorption of reactive power

If $V = V\angle\alpha$ and $I = I\angle\beta$

$$\text{Apparent power } (S) = \text{Real power } (P) + j \text{ Reactive power } (Q)$$
$$S = VI^* = VIe^{j(\alpha + \beta)} = VI\cos(\alpha - \beta) + j\sin(\alpha - \beta)$$
$$\Rightarrow \qquad P + jQ = VI\cos\phi + jVI\sin\phi$$
$$P = VI\cos\phi$$
$$Q = VI\sin\phi$$

Reactive power (Q) is positive when $\varphi = \alpha - \beta > 0$ or I lags V and negative when I leads V. This is the convention used. For this reason, S is taken as VI^* and not V^*I (conjugate is necessary to obtain phase angle difference between V and I).

Figure 2.4 indicates the direction of real and reactive power flow under different operating conditions.

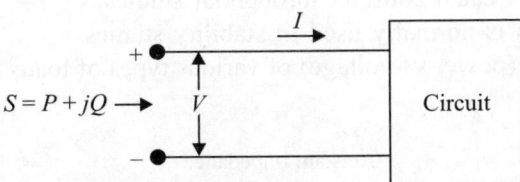

Figure 2.4   Direction of P and Q flow.

The direction of reactive power flow under different operating conditions.

if $P > 0$, circuit absorbs real power
if $P < 0$, circuit generates real power
if $Q > 0$, circuit absorbs reactive power (I lags V)
if $Q < 0$, circuit generates reactive power (I leads V)

When current I lags voltage V by an angle φ between 0 and 90°, we find that $P = VI\cos\phi$ and $Q = VI\sin\phi$ are both positive, indicating real and reactive power absorption by the inductive circuit. When I leads V by an angle φ between 0 and 90°, P is still positive but $Q = VI \sin \phi$ is negative indicating that negative vars are being absorbed or positive vars are being generated by the capacitive circuit.

The static load may be represented in three different ways. They are

- Constant power

$$S = P + jQ \tag{2.13}$$

The MW and MVAR of load are treated constant. This representation is normally used in power flow studies.

- Constant current

In per-unit

$$I = \frac{S^*}{V^*} = \frac{P - jQ}{V^*} = I \angle (\alpha - \phi) \tag{2.14}$$

where, $V = V \angle \alpha$ and $\phi = \tan^{-1} \dfrac{Q}{P}$

Magnitude of current is treated constant in studies (stability).

- Constant impedance

$$Z = \frac{V}{I} = \frac{VV^*}{P - jQ} = \frac{V^2}{P - jQ} = \frac{V^2(P + jQ)}{P^2 + Q^2} = R + jX \tag{2.15}$$

where

$$R = \frac{V^2 P}{P^2 + Q^2} \tag{2.16}$$

and

$$X = \frac{V^2 Q}{P^2 + Q^2} \tag{2.17}$$

This impedance is treated constant throughout studies.

This representation is normally used in stability studies.

The characteristics (power vs voltage) of various types of loads are shown in Figure 2.5.

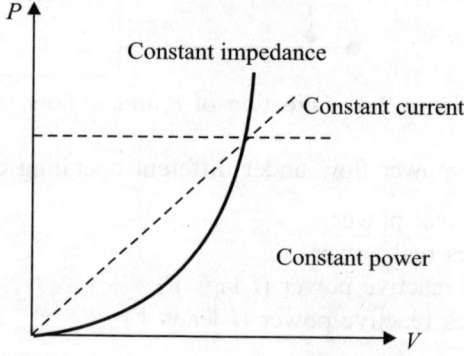

Figure 2.5   Characteristics of various types of loads.

2.1.5   Modelling of Transmission Line

The modelling of transmission line depends on the length of the line.

   (i) Short line is represented by series impedance only.
   (ii) Medium line is represented by π-equivalent (Figure 2.6).
   (iii) Long line is represented by ABCD parameters in the form of π-equivalent.

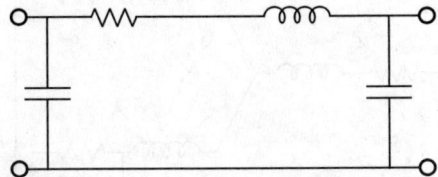

Figure 2.6   Transmission line model.

Series resistance and shunt admittances are neglected in fault and stability studies.

2.1.6   Modelling of Two-winding Transformer

The two-winding transformer is represented by T-equivalent with primary and secondary in series, with magnetising circuit as shunt as shown in Figure 2.7.

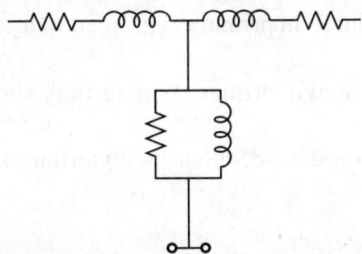

Figure 2.7   Two-winding transformer model.

Series resistance and shunt admittances are neglected in fault and stability studies.

2.1.7   Modelling of Three-winding Transformer

In a three-winding transformer the third winding, usually called *tertiary winding* connected in delta, is used for one or more of the following purposes:

   (i) To give supply to substation auxiliary devices.
   (ii) To connect VAR supply devices.
   (iii) To reduce harmonics by allowing third harmonic current to circulate within the delta windings.
   (iv) To improve waveform and reduce harmonics in HVDC (High Voltage Direct Current) converters.

The three windings may have different MVA ratings. The impedance of each winding given in pu on its own MVA rating may have to be converted to a common base for representing in impedance or reactance diagram.

The equivalent circuit of a three-winding transformer can be represented by the single-phase equivalent circuit as shown in Figure 2.8.

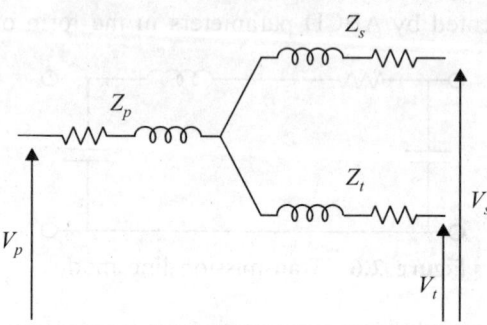

Figure 2.8 Single-phase equivalent of a three-winding transformer.

In the equivalent circuit the impedance per phase of the three windings (referred to one of the windings, usually primary) are connected in star. The magnetising circuit is neglected for simplicity. The three impedances are measured by standard short-circuit tests as seen from Figure 2.8,

Z_{ps} = leakage impedance measured in primary with secondary short-circuited and tertiary open

$\quad = Z_p + Z_s$ (2.18)

Z_{pt} = leakage impedance measured in primary with tertiary short-circuited and secondary open

$\quad = Z_p + Z_t$ (2.19)

Z_{st} = leakage impedance measured in secondary with tertiary short-circuited and primary open

$\quad = Z_s + Z_t$ (2.20)

From Eqs. (2.18) to (2.20), we get

$$Z_p = \frac{(Z_{ps} + Z_{pt} - Z_{st})}{2}$$ (2.21)

$$Z_s = \frac{(Z_{ps} + Z_{st} - Z_{pt})}{2}$$ (2.22)

$$Z_t = \frac{(Z_{pt} + Z_{st} - Z_{ps})}{2}$$ (2.23)

Series resistances are neglected in fault and stability studies.

2.1.8 Modelling of Shunt Capacitor

Let V be the voltage across the shunt capacitor and I be the current entering the capacitor as shown in Figure 2.9. Then in per-unit

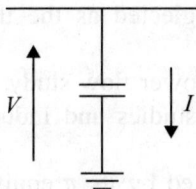

Figure 2.9 Shunt capacitor.

$$S^* = V^*I = -jQ \qquad \text{(since } P = 0\text{)} \qquad (2.24)$$

$$Y = \frac{I}{V} = \frac{-jQ}{VV^*} = \frac{-jQ}{V^2} = jB$$

It may be noted that Q is taken negative for leading reactive power. Shunt capacitor is neglected in fault and stability studies.

2.1.9 Modelling of Shunt Reactor

Let V be the voltage across the shunt reactor and I be the current entering the reactor as shown in Figure 2.10. Then in per-unit

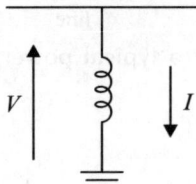

Figure 2.10 Shunt reactor.

$$S^* = V^*I = -jQ$$

$$Y = \frac{I}{V} = \frac{-jQ}{VV^*} = \frac{-jQ}{V^2} = -jB \qquad (2.25)$$

It may be noted that Q is taken positive for lagging reactive power. Shunt reactor is neglected in fault and stability studies.

2.2 Impedance Diagram

The impedance diagram is the equivalent circuit of power system in which the various components are represented by their approximate or simplified equivalent circuits. The impedance diagram is used for power flow studies. The impedance diagram is a positive sequence diagram and drawn from the one-line diagram which is the single-phase equivalent circuit. The following approximations are made while forming the impedance diagram:

(i) The neutral impedance is neglected as the three-phase system is balanced and no neutral current flows.

(ii) All loads are represented in power flow study, but only rotating loads are represented in short circuit and stability studies and induction motor loads are considered only for subtransient study.

(iii) Transmission line is represented by its π equivalent.

(iv) Transformer is represented by T-equivalent (primary, secondary and magnetising circuit).

Impedance diagram for a typical power system having two generators supplying local passive load and through step-up transformer, transmission line, step-down transformer, a motor load and a passive load is shown in Figure 2.11.

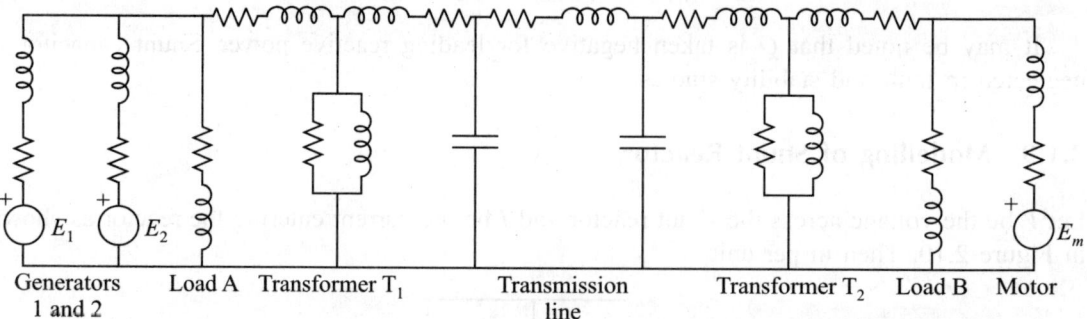

Generators Load A Transformer T_1 Transmission   Transformer T_2 Load B Motor
1 and 2   line

Figure 2.11 Impedance diagram for a typical power system corresponding to Figure 1.8.

2.3 Reactance Diagram

The reactance diagram is the equivalent circuit of power system in which the various components are represented by their reactances. The reactance diagram is used for short circuit and stability studies. The reactance diagram is a positive sequence diagram and drawn from the one-line diagram which is the single-phase equivalent circuit or from the impedance diagram. The following components are neglected while forming the reactance diagram:

(i) The neutral impedance as the three-phase system is balanced and no neutral current flows

(ii) All static loads

(iii) The resistances

(iv) Shunt branches (magnetising circuit of transformers, capacitance of transmission lines and shunt capacitors/reactors)

Reactance diagram for a typical power system having two generators supplying local passive load and through step-up transformer, transmission line and step-down transformer, a motor load and a passive load is shown in Figure 2.12.

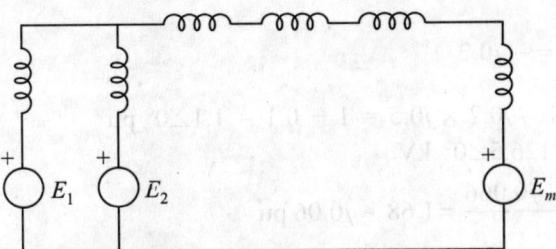

Figure 2.12   Reactance diagram for a typical power system corresponding to Figure 1.8.

Note that both impedance and reactance diagrams are positive sequence diagrams.

EXAMPLE 2.1   The one-line diagram of a three-phase system is shown in Figure 2.13. The transformer reactance is 20% on a base of 100 MVA, 23/115 kV and the line impedance is $Z = j66.125 \ \Omega$. The load at bus 2 is $S_2 = 184.8$ MW $+ j6.6$ MVAR and at bus 3 is $S_3 = 0$ MW $+ j20$ MVAR. It is required to hold the voltage at bus 3 at $115\angle0°$ kV.

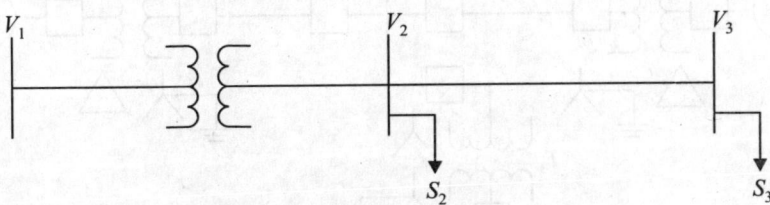

Figure 2.13   One-line diagram for Example 2.1.

Working in per-unit, determine the voltage at buses 2 and 1.

Solution   Let the base values be

$S_b = 100$ MVA and $V_b = 115$ kV on line side

$V_{b3} = V_{b2} = 115$ kV and

$V_{b1} = 115 \times \dfrac{23}{115} = 23$ kV as per transformation

$V_3 = \dfrac{115\angle0°}{115} = 1\angle0°$ pu

$S_3 = \dfrac{0 + j20}{100} = j0.2$ pu

$S_2 = \dfrac{184.8 + j6.6}{100} = 1.848 + j0.066$ pu

$Z_l = j66.125 \times \dfrac{100}{(115)^2} = j0.5$ pu

$Z_T = \dfrac{j20}{100} = j0.2$ pu

$$I_3 = \frac{S_3^*}{V_3^*} = \frac{-j0.2}{1} = -j0.2$$

$$V_2 = V_3 + I_3Z_1 = 1 - j0.2 \times j0.5 = 1 + 0.1 = 1.1\angle 0° \text{ pu}$$
$$= 1.1 \times 115 = 126.5\angle 0° \text{ kV}$$

$$I_2 = \frac{S_2^*}{V_2^*} = \frac{1.848 - j0.066}{1.1} = 1.68 - j0.06 \text{ pu}$$

$$I_1 = I_3 + I_2 = -j0.2 + 1.68 - j0.06 = 1.68 - j0.26$$
$$V_1 = V_2 + I_1 \times Z_T = 1.1 + [(1.68 - j0.26) \times j0.2]$$
$$= 1.1 + 0.052 + j0.336 = 1.152 + j0.336 = 1.2\angle 22.42° \text{ pu}$$
$$= 1.2 \times 23 = 27.6\angle 22.42° \text{ pu}$$

EXAMPLE 2.2 The one-line diagram of an unloaded power system is shown in Figure 2.14.

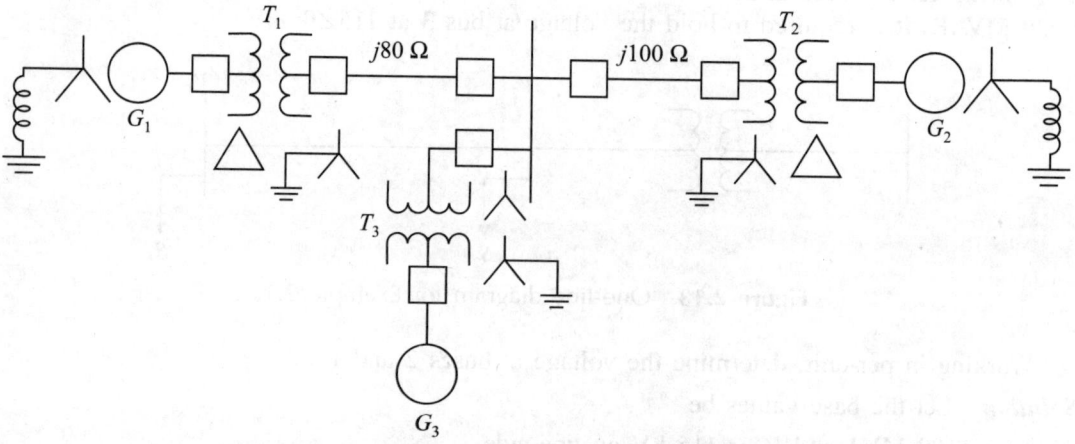

Figure 2.14 One-line diagram for Example 2.2.

Reactances of the two sections of the transmission line are shown in Figure 2.14. The generators and transformers are rated as follows:

Generator (G_1) : 20 MVA, 13.8 kV, $X_d'' = 0.2$ pu
Generator (G_2) : 30 MVA, 18 kV, $X_d'' = 0.2$ pu
Generator (G_3) : 30 MVA, 20 kV, $X_d'' = 0.2$ pu
Transformer (T_1) : 25 MVA, 220 Y/13.8 Δ kV, $X = 10\%$
Transformer (T_2) : Single-phase units, each rated 10 MVA, 127/18 kV, $X = 10\%$
Transformer (T_3) : 35 MVA, 220 Y/22Y kV, $X = 10\%$

Draw the reactance diagram with all reactances marked in pu. Choose a base of 50 MVA, 13.8 kV in the circuit of generator G_1.

Solution

MVA base   = 50

Voltage base = 13.8 kV in the circuit of G_1

$$= 220 \text{ kV in the line as per transformation by } T_1 \left(\frac{V_{G_1}}{V_{b_l}} = \frac{13.8}{220} \right)$$

$$= 18 \text{ kV in } G_2 \text{ as per transformation by } T_2 \left(\frac{V_{G_2}}{V_{b_l}} = \frac{18}{127\sqrt{3}} \right)$$

$$= 22 \text{ kV in } G_3 \text{ as per transformation by } T_3.$$

∴ The pu reactances of various components are:

Transmission line 1 : $80 \times \dfrac{50}{220^2}$ = 0.0826 pu

Transmission line 2 : $100 \times \dfrac{50}{(127\sqrt{3})^2}$ = 0.1033 pu

Generator (G_1)   : $0.2 \times \left(\dfrac{13.8}{13.8}\right)^2 \times \dfrac{50}{20}$ = 0.5 pu

Generator (G_2)   : $0.2 \times \left(\dfrac{18}{18}\right)^2 \times \dfrac{50}{30}$ = 0.3333 pu

Generator (G_3)   : $0.2 \times \left(\dfrac{20}{22}\right)^2 \times \dfrac{50}{30}$ = 0.2755 pu

Transformer (T_1)   : $0.1 \times \left(\dfrac{220}{220}\right)^2 \times \dfrac{50}{25}$ = 0.2 pu

Transformer (T_2)   : $0.1 \times \left(\dfrac{127\sqrt{3}}{220}\right)^2 \times \dfrac{50}{30}$ = 0.166 pu

Transformer (T_3)   : $0.1 \times \left(\dfrac{220}{220}\right)^2 \times \dfrac{50}{35}$ = 0.1429 pu

The impedance diagram is shown in Figure 2.15.

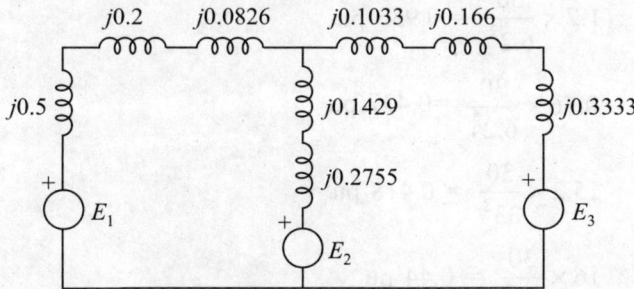

Figure 2.15   Impedance diagram of system of Figure 2.14.

EXAMPLE 2.3   Obtain the pu impedance diagram of the power system shown in Figure 2.16.

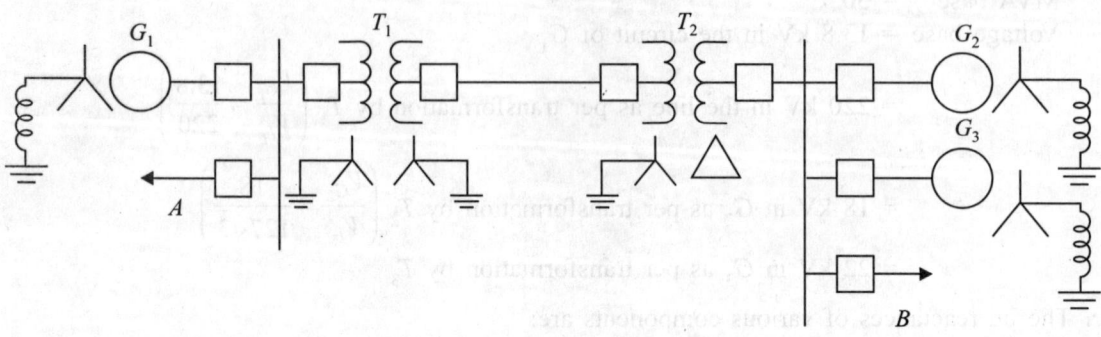

Figure 2.16   One-line representation of a power system.

Generator (G_1)	: 30 MVA, 10.5 kV, $X'' = 1.6\ \Omega$
Generator (G_2)	: 15 MVA, 6.6 kV, $X'' = 1.2\ \Omega$
Generator (G_3)	: 25 MVA, 6.6 kV, $X'' = 0.56\ \Omega$
3-ph Transformer (T_1)	: 15 MVA, 33/11 kV, $X = 15.2\ \Omega$ per phase on high tension side
3-ph Transformer (T_2)	: 15 MVA, 33/6.2 kV, $X = 16\ \Omega$ per phase on high tension side
Transmission line	: 20.5 Ω/phase
Load A	: 15 MW, 11 kV, 0.9 pf lagging
Load B	: 40 MW, 6.6 kV, 0.85 pf lagging

Solution   Choose a common three-phase MVA base of 30, and a voltage base of 33 kV line-to-line on the transmission line. Then the base voltage of

Generator (G_1), Load A : 11 kV (line-to-line) as per transformation by T_1
Generator (G_2), (G_3), Load B : 6.2 kV (line-to-line) as per transformation by T_2

The pu reactances of various components are:

Transmission line: $20.5 \times \dfrac{30}{33^2} = 0.564$ pu

Generator (G_1)   : $1.6 \times \dfrac{30}{11^2} = 0.396$ pu

Generator (G_2)   : $1.2 \times \dfrac{30}{6.2^2} = 0.936$ pu

Generator (G_3)   : $0.56 \times \dfrac{30}{6.2^2} = 0.437$ pu

Transformer (T_1) : $15.2 \times \dfrac{30}{33^2} = 0.418$ pu

Transformer (T_2) : $16 \times \dfrac{30}{33^2} = 0.44$ pu

Series representation of load in per-unit

$$S^* = V^*I = V^*\frac{V}{Z} = \frac{V^2}{Z} \quad \text{or} \quad Z = \frac{V^2}{S^*}$$

$$Z_A = \frac{\left(\dfrac{11}{11}\right)^2}{\dfrac{(15 - j15\cos^{-1} 0.9)}{30}}$$

$$= \frac{1}{0.5 - j0.2422}$$

$$= 1.62 + j0.875$$

Similarly,

$$Z_B = \frac{\left(\dfrac{6.6}{6.2}\right)^2}{\dfrac{(40 - j40\cos^{-1} 0.85)}{30}}$$

$$= 0.614 + j0.381$$

The impedance diagram of the system is shown in Figure 2.17.

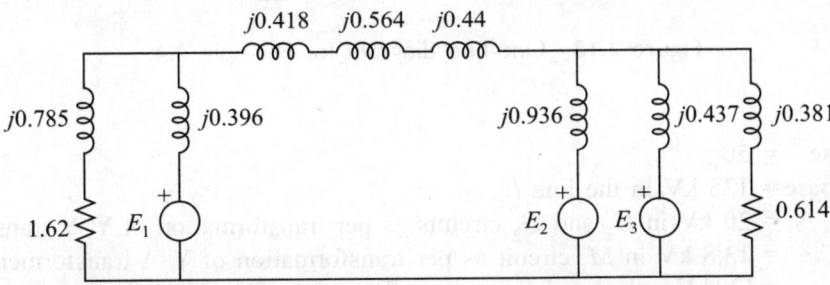

Figure 2.17 Impedance diagram of the system of Figure 2.16.

EXAMPLE 2.4 Draw the impedance diagram for the power system shown in Figure 2.18. Mark impedances in pu. Neglect resistance and use a base of 50 MVA, 138 kV in the 40 Ω line. The ratings of the generators, motors and transformers are

Generator (G_1)   : 20 MVA, 18 kV, X_d'' = 20%
Generator (G_2)   : 20 MVA, 18 kV, X_d'' = 20%
Synchronous motor (M_3) : 30 MVA, 13.8 kV, X_d'' = 20%
3-ph Y–Y transformer   : 20 MVA, 138 Y/20 Y kV, X = 10%
3-ph Y–Δ transformer   : 15 MVA, 138 Y/13.8Δ kV, X = 10%

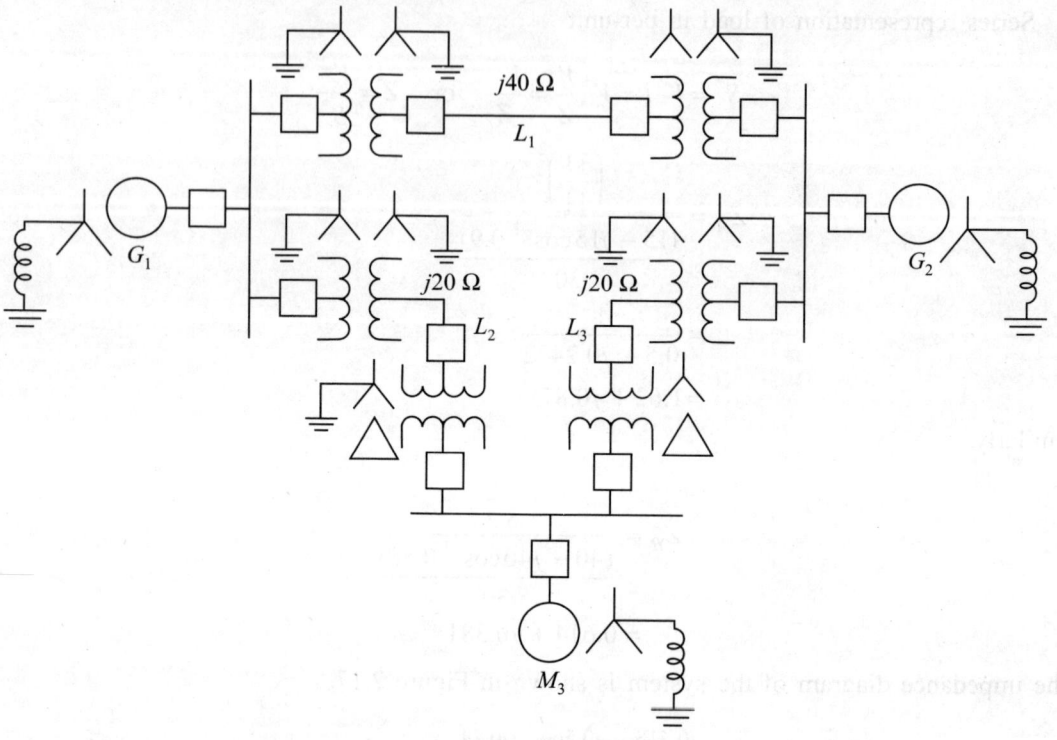

Figure 2.18 One-line diagram for Example 2.4.

Solution

MVA base $= 50$

Voltage base $= 138$ kV in the line L_1.

$\quad\quad\quad\quad\quad = 20$ kV in G_1 and G_2 circuits as per transformation of Y–Y transformers.

$\quad\quad\quad\quad\quad = 13.8$ kV in M_3 circuit as per transformation of Y–Δ transformers.

$\quad\quad\quad\quad\quad = 138$ kV in the line L_2 as per transformation of Y–Y transformers from G_1 circuit.

$\quad\quad\quad\quad\quad = 138$ kV in the line L_3 as per transformation of Y–Y transformers from G_2 circuit.

$\therefore$ The pu reactances of various components are

Transmission line 1 $\quad : 40 \times \dfrac{50}{138^2}$ $\quad\quad\quad\quad = 0.105$ pu

Transmission line 2, 3 $: 20 \times \dfrac{50}{138^2}$ $\quad\quad\quad\quad = 0.0525$ pu

Generator (G_1) $\quad\quad\quad : 0.2 \times \left(\dfrac{18}{20}\right)^2 \times \dfrac{50}{20} = 0.405$ pu

Generator (G_2) $\quad\quad\quad : 0.2 \times \left(\dfrac{18}{20}\right)^2 \times \dfrac{50}{20} = 0.405$ pu

Motor (M_3) $\qquad : 0.2 \times \left(\dfrac{13.8}{13.8}\right)^2 \times \dfrac{50}{30} = 0.3333$ pu

Y–Y transformer $\qquad : 0.1 \times \left(\dfrac{138}{138}\right)^2 \times \dfrac{50}{20} = 0.25$ pu

Y–Δ transformer $\qquad : 0.1 \times \left(\dfrac{138}{138}\right)^2 \times \dfrac{50}{15} = 0.33$ pu

The impedance diagram of the system is shown in Figure 2.19.

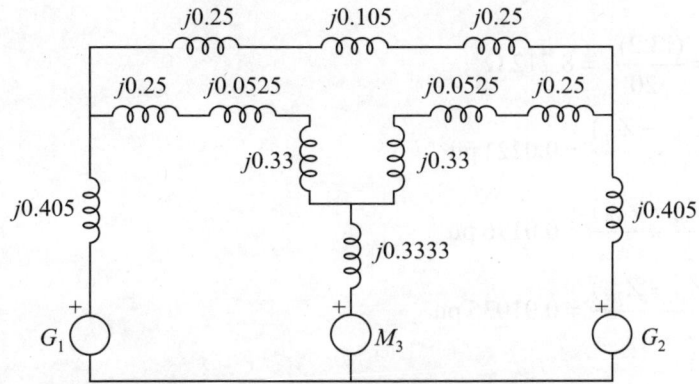

Figure 2.19 Impedance diagram for system shown in Figure 2.18.

EXAMPLE 2.5 The three-phase ratings of a three-winding transformer are:

Primary $\quad$: Star-connected, 110 kV, 20 MVA
Secondary : Star-connected, 13.2 kV, 15 MVA
Tertiary $\quad$: Delta-connected, 2.1 kV, 0.5 MVA

Three short circuit tests performed on this transformer yielded the following results:

Test 1 $\quad$ Primary excited, secondary shorted: 2290 V, 52.5 A
Test 2 $\quad$ Primary excited, tertiary shorted: 1785 V, 52.5 A
Test 3 $\quad$ Secondary excited, tertiary shorted: 148 V, 328 A

Find the pu impedances of the star-connected, single-phase equivalent circuit for a base of 20 MVA, 110 kV in primary circuit.

Solution: Tests 1 and 2 are performed on primary winding and therefore, the pu impedances of Z_{ps} and Z_{pt} can be directly obtained using the primary winding ratings as base values. Test 3 is performed on secondary winding, hence the pu impedance of Z_{st} is obtained using the secondary winding voltage rating as base.

Base MVA, $S_b = 20$ MVA

Base impedance of primary circuit

$$Z_{bp} = \frac{(V_{bp})^2}{S_b} = \frac{110^2}{20} = 605 \ \Omega$$

$$Z_{ps} = \frac{2290/\sqrt{3}}{52.5} = 25.1835 \ \Omega = \frac{25.1835}{605} = 0.0416 \ \text{pu}$$

$$Z_{pt} = \frac{1785/\sqrt{3}}{52.5} = 19.6299 \ \Omega = \frac{19.6299}{605} = 0.0324 \ \text{pu}$$

Base impedance of secondary circuit

$$Z_{bs} = \frac{(V_{bs})^2}{S_b} = \frac{(13.2)^2}{20} = 8.712 \ \Omega$$

$$Z_p = \frac{(Z_{ps} + Z_{pt} - Z_{st})}{2} = 0.0221 \ \text{pu}$$

$$Z_s = \frac{(Z_{ps} + Z_{st} - Z_{pt})}{2} = 0.0196 \ \text{pu}$$

$$Z_t = \frac{(Z_{pt} + Z_{st} - Z_{ps})}{2} = 0.01035 \ \text{pu}$$

Short Answer Questions

1. What is an impedance diagram and what are the approximations made in this diagram?
   The impedance diagram is the equivalent circuit of power system in which the various components are represented by their approximate or simplified equivalent circuits. The following approximations are made while forming the impedance diagram:
   (i) The neutral impedance is neglected.
   (ii) All static loads are represented as impedances.
   (iii) Transmission line is represented by its π equivalent.

2. What is a reactance diagram and what are the approximations made in this diagram?
   The reactance diagram is the equivalent circuit of power system in which the various components are represented by their reactances. The following components are neglected while forming the reactance diagram:
   (i) The neutral impedance
   (ii) All static loads
   (iii) The resistances
   (iv) Shunt branches (magnetising circuit of transformers capacitance of transmission lines and shunt capacitors/reactors)

3. How are the loads represented in impedance or reactance diagram?

In impedance diagram, the load is modelled as resistance in series with reactance as:

$$R = \frac{V^2 P}{P^2 + Q^2} \quad \text{and} \quad X = \frac{V^2 Q}{P^2 + Q^2}$$

4. What is the step-by-step procedure to be followed to draw the per-unit impedance/reactance diagram of a power system?

Step 1   Choose the three-phase MVA of the largest section as common base MVA.

Step 2   Choose the line-to-line kV of the largest section as base kV and calculate the base kV of other sections depending on the line-to-line voltage ratio of the transformers connecting them. Base MVA remains same in all parts of the system.

Step 3   Calculate per-unit impedance in each section. Impedance information of components is usually given in per-unit or per cent based on its own ratings and change this to base values chosen. If impedance value is given in ohm, pu impedance is calculated from the base values chosen.

Step 4   For three single-phase transformers connected as a three-phase unit, three-phase ratings are determined from single-phase ratings as

(a) Base MVA = MVA_{3ph} = 3 × MVA_{1ph}

(b) Base voltage = kV_l = $\sqrt{3}\ kV_{ph}$ for star connected side = kV_{ph} for delta connected side

(c) The pu impedance of three-phase unit is the same as that of the single-phase unit.

Step 5   Draw the one-line diagram.

Step 6   Draw the impedance/reactance diagram from this one-line diagram.

5. What are the different load models?

The loads are modelled as

Constant power: $S = P + jQ$

MW and MVAR of load are treated constant. This representation is normally used in power flow studies.

Constant current: In per-unit

$$I = \frac{S^*}{V^*} = \frac{P - jQ}{V^*} = I\angle(\alpha - \phi)$$

where $V = V\angle\alpha$ and $\phi = \tan^{-1}\frac{Q}{P}$

Magnitude of current is treated constant in studies (stability).

Constant impedance:

$$Z = \frac{V}{I} = \frac{VV^*}{P - jQ} = \frac{V^2}{P - jQ} = \frac{V^2 P + jQ}{P^2 + Q^2} = R + jX$$

where, $R = \frac{V^2 P}{P^2 + Q^2}$ and $X = \frac{V^2 Q}{P^2 + Q^2}$

This impedance is treated constant throughout the studies.

Exercises

2.1 Draw an impedance diagram for the electric power system shown in Figure E2.1. Choose 100 MVA and 20 kV on generator side as the bases. The three-phase power, line-to-line voltage ratings and impedances in per cent are given:

G_1 = 90 MVA, 20 kV, X = 9%
T_1 = 80 MVA, 20/200 kV, X = 16%
T_2 = 80 MVA, 200/20 kV, X = 20%
G_2 = 90 MVA, 18 kV, X = 9%
Line = 200 kV, X = 120 Ω
Load = 200 kV, S = 48 MW + j64 MVAR

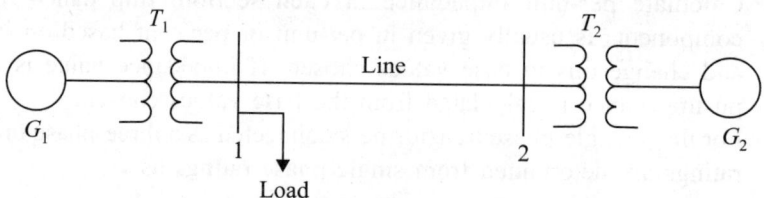

Figure E2.1 One-line diagram for Problem 2.1.

[**Ans.** j0.1, j0.2, j0.25, j0.081, j0.3, 0.75 + j1.0]

2.2 The ratings of the three-winding three-phase transformer are

Primary   = Y-connected, 40 MVA, 110 kV
Secondary = Y-connected, 40 MVA, 22 kV
Tertiary = Δ-connected, 15 MVA, 4 kV

The per phase measured reactances at the terminal of the first winding with the second one short-circuited and the third open circuited are

Z_{ps} = 9.6%, 40 MVA, 110/22 kV
Z_{pt} = 7.2%, 40 MVA, 110/4 kV
Z_{st} = 12%, 40 MVA, 22/4 kV

Obtain the T equivalent impedances of the three-winding transformer to the common 100 MVA base.   [**Ans.** Z_p = j0.06, Z_s = 0.18, Z_t = j0.12]

2.3 The three-phase power and line-to-line ratings of the electric power system shown in Figure 2.2 are given:

G_1 = 60 MVA, 20 kV, X = 9%
T_1 = 50 MVA, 20/200 kV, X = 10%
T_2 = 50 MVA, 200/20 kV, X = 10%
M = 43.2 MVA, 18 kV, X = 8%
Line = 200 kV, Z = 120 + j200 Ω

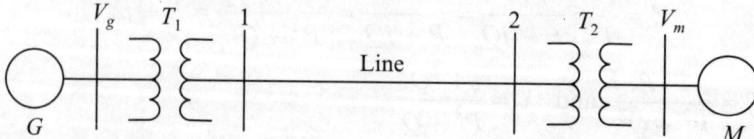

Figure E2.2 One-line diagram for Problem 2.3.

(i) Draw an impedance diagram showing all impedances in per-unit on a 100 MVA base. Choose 20 kV as the voltage base for generator.

(ii) The motor is drawing 45 MVA, 0.80 power factor lagging at line-to-line terminal voltage of 18 kV. Determine the terminal voltage and the internal emf of the generator in kV.

[**Ans.** (i) $j0.15$, $j0.2$, $j0.2$, $j0.15$, $0.3 + j0.5$, (ii) 26.36 kV, 27.5 kV]

2.4 The one-line diagram of a three-phase power system is shown in Figure E2.3. Impedances are marked in pu on a 100 MVA, 400 kV base. The load at bus 2 is $S_2 = 15.93$ MW $- j33.4$ MVAR and at bus 3 is $S_3 = 77$ MW $+ j14$ MVAR.

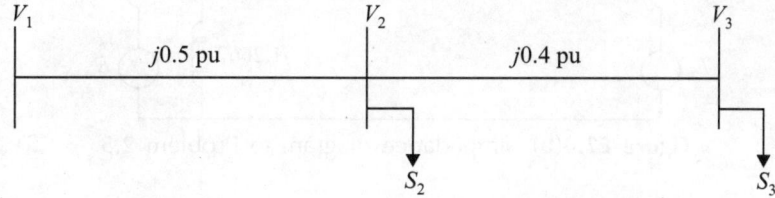

Figure E2.3 One-line diagram for Problem 2.4.

It is required to hold the voltage at bus 3 at $400\angle0°$ kV. Working in per-unit, determine the voltage at buses 2 and 1. [**Ans.** 440 kV, 480 kV]

2.5 The one-line diagram of the three-phase power system is shown in Figure E2.4. Select a common base of 100 MVA and 22 kV on the generator side. Draw an impedance diagram with all impedances including the load impedances marked in per-unit. The manufacturer's data for each device is given:

$G = 90$ MVA, 22 kV, $X = 18\%$
$T_1 = 50$ MVA, 22/220 kV, $X = 10\%$
$T_2 = 40$ MVA, 220/11 kV, $X = 6\%$
$T_3 = 40$ MVA, 22/110 kV, $X = 6.4\%$
$T_4 = 40$ MVA, 110/11 kV, $X = 8\%$
$M = 66.5$ MVA, 10.45 kV, $X = 18.5\%$

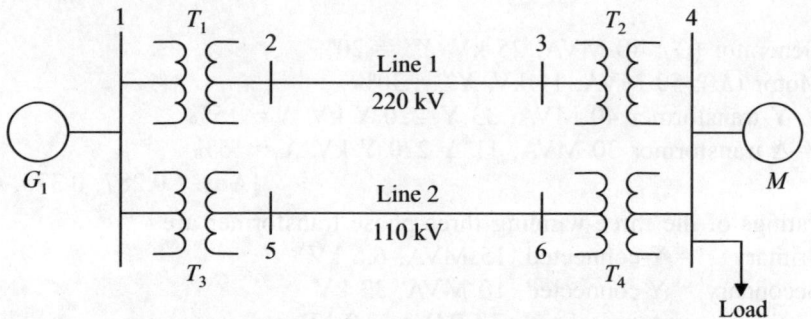

Figure E2.4(a) One-line diagram for Problem 2.5.

The three-phase load at bus 4 absorbs 57 MVA, 0.6 power factor lagging at 10.45 kV. Line 1 and Line 2 have reactances of 48.4 Ω and 65.43 Ω respectively.

[**Ans.** The impedance diagram is shown in Figure E2.4(b)]

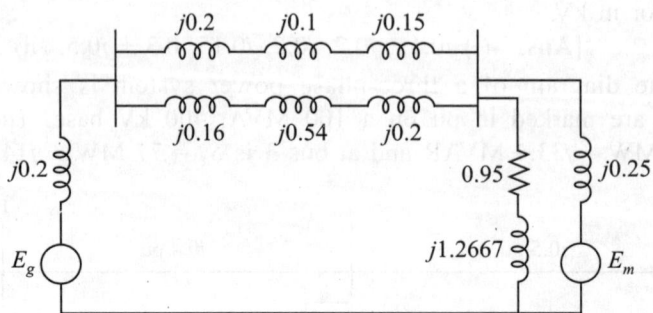

Figure E2.4(b) Impedance diagram to Problem 2.5.

2.6 The motor of Problem 2.5 operates at full load 0.8 power factor leading at a terminal voltage of 10.45 kV.
   (i) Determine the voltage at the generator bus bar (bus 1)
   (ii) Determine the generator and motor internal emfs.
                    [**Ans.** $22.09\angle15.95°$ kV, $23.82\angle25.14°$ kV, $11.7\angle-7.56°$ kV]

2.7 Draw the impedance diagram for the power system shown in Figure E2.5. Use a base of 100 MVA, 220 kV in 50 Ω line. The ratings of the generator, motor and transformers are

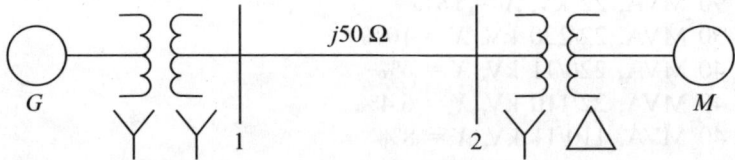

Figure E2.5 Schematic diagram for Problem 2.7.

Generator (G): 40 MVA, 25 kV, $X'' = 20\%$
Motor (M): 50 MVA, 11 kV, $X'' = 30\%$
Y–Y transformer 40 MVA, 33 Y–220 Y kV, $X = 15\%$
Y–Δ transformer 30 MVA, 11 Δ–220 Y kV, $X = 15\%$

[**Ans.**   0.287, 0.375, 0.1, 0.5, 0.6]

2.8 The ratings of the three-winding three-phase transformer are
   Primary   = Y-connected, 15 MVA, 6.6 kV
   Secondary = Y-connected, 10 MVA, 33 kV
   Tertiary   = Δ-connected, 7.5 MVA, 2.2 kV

With resistance neglected the following leakage impedances are calculated from short circuit tests:

Measured from primary side $Z_{ps} = j0.232\ \Omega$, $Z_{pt} = j0.29\ \Omega$.

Measured from secondary side $Z_{st} = 8.7\ \Omega$.

Find the impedances of the Y-connected equivalent circuit on a base of 15 MVA, 6.6 kV in primary circuit. 　　　　　　　　　　　　　　　　**[Ans.** $j0.03$, $j0.05$, $j0.07$ pu]

CHAPTER 3

Power Flow Analysis

Power flow analysis or study is required for planning, operation and control of existing system as well as for planning of future expansion. The economical operation of the existing system depends upon knowing the effect of generation dispatch on the loading of transmission lines and transformers, and the voltage conditions of various buses of the power system. The planning of optimum location and size of new generating sources (active/reactive), and interconnections are also based on this information, which are provided by a number of power flow studies carried out considering various options under both normal and abnormal system conditions. The power flow analysis also gives the initial conditions of the power system for carrying out short-circuit and stability studies.

3.1 Power Flow Problem

The power flow problem is the calculation of voltage magnitude and phase angle of the buses in the power system. Once these quantities are known, active and reactive power flow over lines and transformers, injection in to or drawl from buses can be computed.

3.2 Types of Buses

Each bus in a power system is associated with the following four quantities:

- Magnitude of voltage, $|V|$
- Phase angle of voltage, δ
- Active power, P
- Reactive power, Q

In a power flow problem, two out of the four quantities are specified and the remaining two are determined by the solution of power flow equations. Accordingly, the buses are classified into three types, namely load bus, generation or voltage controlled bus and slack or swing or reference bus. The quantities specified and to be determined for the different types of buses are given in Table 3.1.

Table 3.1   Classification of buses

Bus	Specified quantities	Quantities to be determined
Load bus, PQ bus	P, Q	$\lvert V \rvert, \delta$
Generation bus, PV bus	$P, \lvert V \rvert$	Q, δ
Slack bus, $V\delta$ bus	$\lvert V \rvert, \delta$	P, Q

3.3   Need and Selection of Slack Bus

In a power system generation and load are known, *apriori*, but transmission and transformation losses will be known only after the power flow solution is completed because losses depend on the relative location and dispatch of generation with respect to load. It is, therefore, required to consider one of the generation buses as slack bus which will supply the power losses (active and reactive).

The slack bus is selected based on the following considerations:

- Maximum power
- Location near centre of gravity of load
- Costly generation

3.4   Bus Admittance Matrix

The power generated in the various generating stations flow through the transmission network on its way to the different load centres. The modelling of network, therefore, becomes necessary for power flow studies.

The network model depends on the method used for power flow solution—node/bus method or loop/mesh method. The node method or bus frame of analysis, which makes use of bus admittance matrix, is more suitable for large network and is used for power flow studies.

Bus admittance matrix can be computed by inspection (when there is no mutual coupling between lines), singular transformation and inversion of bus impedance matrix.

In power flow analysis, no mutual coupling is assumed between transmission lines and therefore the formation of bus admittance matrix by inspection method, which is simple, is normally used.

3.4.1   Formation of Bus Admittance Matrix by Inspection Method

First a three-bus system is considered as shown in Figure 3.1 for formulation of bus admittance matrix Y_{bus} and is generalized later for n-bus system.

At bus 1,

$$
\begin{aligned}
I_1 &= I_{11} + I_{12} + I_{13} \\
&= V_1 y_{11} + (V_1 - V_2)y_{12} + (V_1 - V_3)y_{13} \\
&= V_1(y_{11} + y_{12} + y_{13}) - V_2 y_{12} - V_3 y_{13} \\
&= V_1 Y_{11} + V_2 Y_{12} + V_3 Y_{13}
\end{aligned}
$$

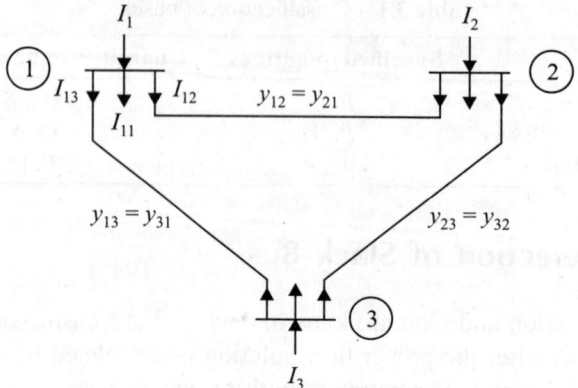

Figure 3.1 A typical three-bus system.

where,

$$Y_{11} = y_{11} + y_{12} + y_{13}$$
$$Y_{12} = -y_{12}$$
$$Y_{13} = -y_{13}$$

Similarly, nodal or bus equations for other buses can also be written. The equations thus obtained in matrix form are

$$\begin{pmatrix} I_1 \\ I_2 \\ I_3 \end{pmatrix} = \begin{pmatrix} Y_{11} & Y_{12} & Y_{13} \\ Y_{21} & Y_{22} & Y_{23} \\ Y_{31} & Y_{32} & Y_{33} \end{pmatrix} \begin{pmatrix} V_1 \\ V_2 \\ V_3 \end{pmatrix}$$

i.e.,

$$I_{bus} = Y_{bus} \, V_{bus} \tag{3.1}$$

and

$$I_p = \sum_{q=1}^{n} Y_{pq} V_q \qquad p = 1, 2, 3$$

For n bus system; matrix form is

$$\begin{pmatrix} I_1 \\ I_2 \\ \vdots \\ I_p \\ \vdots \\ I_n \end{pmatrix} = \begin{pmatrix} Y_{11} & Y_{12} & \cdots & Y_{1p} & \cdots & Y_{1n} \\ Y_{21} & Y_{22} & \cdots & Y_{2p} & \cdots & Y_{2n} \\ & & \vdots & & & \\ Y_{p1} & Y_{p2} & \cdots & Y_{pp} & \cdots & Y_{pn} \\ & & \vdots & & & \\ Y_{n1} & Y_{n2} & \cdots & Y_{np} & \cdots & Y_{nn} \end{pmatrix} \begin{pmatrix} V_1 \\ V_2 \\ \vdots \\ V_p \\ \vdots \\ V_n \end{pmatrix}$$

$$I_p = \sum_{q=1}^{n} Y_{pq} V_q \qquad p = 1, 2, 3, ..., n \tag{3.2}$$

The elements of the bus admittance matrix are computed as follows:

1. The diagonal element corresponding to a particular bus is the sum of all the admittances connected to that bus

$$Y_{pp} = \sum_{q=1}^{n} y_{pq} \qquad (3.3)$$

2. The off-diagonal element is the negative of the admittances connected between the buses

$$Y_{pq} = -y_{pq} \qquad (3.4)$$

Algorithm for building Y_{bus} matrix

Step 1 Get the bus and line data.

Step 2 Compute the diagonal elements of the matrix.

$$Y_{pp} = \sum_{q=1}^{n} y_{pq} \qquad (p = 1, 2, ..., n)$$

Step 3 Compute the off-diagonal elements of the matrix.

$$Y_{pq} = Y_{qp} = -y_{pq} \qquad (p = 1, 2, ..., n \neq q; \; q = 1, 2, ..., n \neq p)$$

The diagonal element Y_{pp} can be found by applying a unit voltage source between the pth bus and the reference bus, and measuring the current flowing into the pth bus, while all other buses are short circuited and is therefore called the *short-circuit driving point* or *self admittance*. Similarly, the off-diagonal element Y_{pq} is found by applying a unit voltage source between the pth bus and the reference bus and measuring current into the qth bus, while all other buses are short circuited and is therefore called *short-circuit transfer* or *mutual admittance*.

$$Y_{pp} = \frac{I_p}{V_p}\bigg|_{V_q=0; \, q \neq p}$$

$$Y_{pq} = Y_{qp} = \frac{I_q}{V_p}\bigg|_{V_q=0; \, q \neq p}$$

3.4.2 Formation of Y_{bus} by Singular Transformation

The Y_{bus} can be formulated by singular transformation given by graph theoretical approach.

Graph

Geometrical features of a network can be described easily if the network is replaced by line segments called *elements* whose terminals are called *nodes*. A linear graph depicts the inter connection of the elements. If each element of a graph is assigned a direction, then the graph is called an *oriented graph*.

In power networks, one node (normally described by 0) is always at ground potential and is considered as reference node. In one-line diagram the neutral of the three-phase balanced system, which is invariably at ground potential is taken as reference and not shown in the diagram. The remaining nodes are the buses either at which power is injected or from which power is drawn. The graph for the typical three-bus system of Figure 3.1 is shown in Figure 3.2. Each source and the shunt admittance connected across it are represented by a single element, called *primitive network* is shown in Figure 3.3.

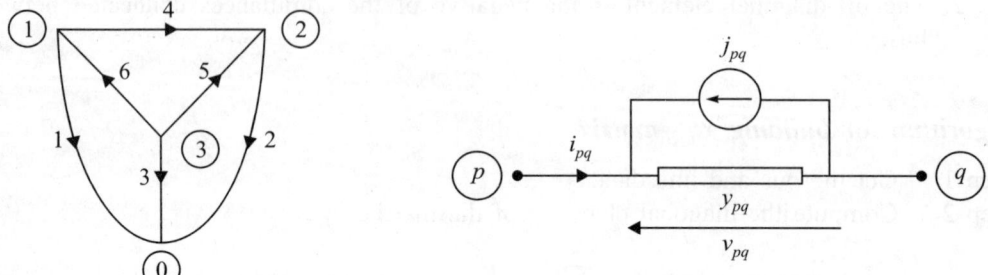

Figure 3.2   Graph for system of Figure 3.1.   **Figure 3.3**   Primitive network.

From Figure 3.3 by Kirchhoff's Current Law (KCL)

$$i_{pq} + j_{pq} = y_{pq}v_{pq} \tag{3.5}$$

For a set of elements

$$I + J = YV \tag{3.6}$$

where, V and I are element voltage and current vectors, J current source vector and Y primitive admittance matrix.

A connected subgraph containing all the nodes of a graph but having no closed path is called a *tree*. The elements of a tree are called *branches*. The number of branches (b) that form a tree is equal to the number of buses (n). Those elements of the graph that are not included in the tree are called *links* (l) and they form a subgraph, not necessarily connected, called *co-tree*. Tree and co-tree are complements of a graph.

$$l = e - b \tag{3.7}$$

where, e is the total number of elements.

A tree and the corresponding co-tree of the graph of Figure 3.2 are shown in Figure 3.4.

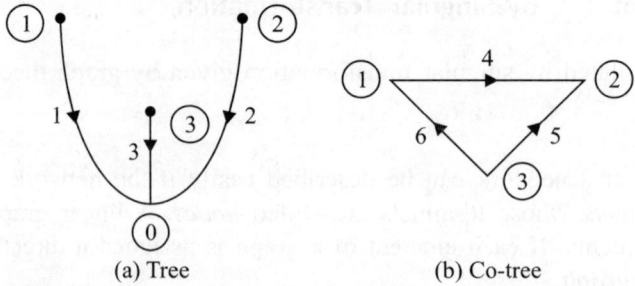

(a) Tree   (b) Co-tree

Figure 3.4   Tree and co-tree of graph of Figure 3.2.

If a link is added to the tree, the graph contains one closed path called *loop*. The number of loops is equal to the number of links.

Bus incidence matrix

The knowledge of b tree voltages is sufficient to compute all element voltages and therefore all bus currents for the given element admittances.

For the system of Figure 3.2, the relations between the six-element voltages and the three-bus voltages V_1, V_2 and V_3 are given below:

$$V_{e1} = V_1$$
$$V_{e2} = V_2$$
$$V_{e3} = V_3$$
$$V_{e4} = V_1 - V_2$$
$$V_{e5} = V_3 - V_2$$
$$V_{e6} = V_3 - V_1$$

In matrix form

$$
\text{Elements} \left\{
\begin{pmatrix} V_{e1} \\ V_{e2} \\ V_{e3} \\ V_{e4} \\ V_{e5} \\ V_{e6} \end{pmatrix}
= \overset{\overbrace{\text{Buses}}}{\begin{matrix} 1 & 2 & 3 \end{matrix}}
\begin{pmatrix} 1 & 0 & 0 \\ 0 & 1 & 0 \\ 0 & 0 & 1 \\ 1 & -1 & 0 \\ 0 & -1 & 1 \\ -1 & 0 & 1 \end{pmatrix}
\begin{pmatrix} V_1 \\ V_2 \\ V_3 \end{pmatrix}
\right.
\tag{3.8}
$$

$$V = AV_{\text{bus}} \tag{3.9}$$

where A is the bus incidence matrix.

$$[A] = \left[\frac{A_b}{A_l} \right] = \left[\frac{U}{A_l} \right]^{\text{branches}}_{\text{links}}$$

where U is unit matrix of size $b \times b$.

The element a_{ik} of the bus incidence matrix connecting nodes i and k is found as per the following rules:

$a_{ik} = 1$ if ith element is incident to and oriented away from kth bus.

$\quad = -1$ if ith element is incident to but oriented towards the kth bus.

$\quad = 0$ if ith element is not incident to the kth bus. $\tag{3.10}$

Substituting Eq. (3.9) into Eq. (3.6), we get

$$I + J = YAV_{\text{bus}}$$

Pre-multiplying by A^T

$$A^T I + A^T J = A^T YA \ V_{\text{bus}} \tag{3.11}$$

Each component of the n-dimensional vector $A^T I$ is the algebraic sum of the element currents leaving the buses 1, 2, ..., n. By application of KCL, we get

$$A^T I = 0$$

Similarly, each component of the vector $A^T J$ is the algebraic sum of all source currents injected into the buses 1, 2, ..., n. These components are therefore the bus currents.

i.e.

$$A^T J = J_{bus}$$

Equation (2.11) is therefore,

$$J_{bus} = A^T Y A\ V_{bus} = Y_{bus} V_{bus}$$

where

$$Y_{bus} = A^T Y A \tag{3.12}$$

Let us find the Y_{bus} using singular transformation for the system of Figure 3.1.

$$Y = \begin{pmatrix} y_{11} & 0 & 0 & 0 & 0 & 0 \\ 0 & y_{22} & 0 & 0 & 0 & 0 \\ 0 & 0 & y_{33} & 0 & 0 & 0 \\ 0 & 0 & 0 & y_{44} & 0 & 0 \\ 0 & 0 & 0 & 0 & y_{55} & 0 \\ 0 & 0 & 0 & 0 & 0 & y_{66} \end{pmatrix}$$

By using A from Eq. (3.9), we get

$$YA = \begin{pmatrix} y_{11} & 0 & 0 \\ 0 & y_{22} & 0 \\ 0 & 0 & y_{33} \\ y_{12} & -y_{12} & 0 \\ 0 & -y_{23} & y_{23} \\ -y_{13} & 0 & y_{13} \end{pmatrix}$$

$$Y_{bus} = A^T YA = \begin{pmatrix} y_{11} + y_{12} + y_{13} & -y_{12} & -y_{13} \\ -y_{12} & y_{22} + y_{12} + y_{13} & -y_{23} \\ -y_{13} & -y_{23} & y_{33} + y_{23} + y_{13} \end{pmatrix}$$

Note that the elements of this matrix agree with those previously calculated using inspection method.

EXAMPLE 3.1 Using singular transformation method, determine Y_{bus} for the network shown in Figure 3.5, where the impedances labelled are shown in per-unit.

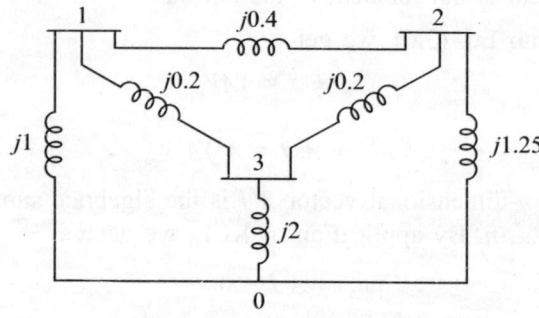

Figure 3.5 Example 3.1.

Solution The oriented graph for the network of Figure 3.5 is the same as shown in Figure 3.2. The corresponding bus incidence matrix obtained using Eq. (3.10) is

$$A = \begin{array}{c} \overbrace{\begin{array}{ccc} 1 & 2 & 3 \end{array}}^{\text{Buses}} \\ \left. \begin{pmatrix} 1 & 0 & 0 \\ 0 & 1 & 0 \\ 0 & 0 & 1 \\ 1 & -1 & 0 \\ 0 & -1 & 1 \\ -1 & 0 & 1 \end{pmatrix} \begin{array}{c} 1 \\ 2 \\ 3 \\ 4 \\ 5 \\ 6 \end{array} \right\} \text{Elements} \end{array}$$

The primitive admittance matrix is

$$Y = \begin{pmatrix} \dfrac{1}{j1} & & & & & \\ & \dfrac{1}{j1.25} & & & & \\ & & \dfrac{1}{j2} & & & \\ & & & \dfrac{1}{j0.4} & & \\ & & & & \dfrac{1}{j0.2} & \\ & & & & & \dfrac{1}{j0.2} \end{pmatrix} = \begin{pmatrix} -j1 & & & & & \\ & -j0.8 & & & & \\ & & -j0.5 & & & \\ & & & -j2.5 & & \\ & & & & -j5 & \\ & & & & & -j5 \end{pmatrix}$$

By using A, we get

$$YA = \begin{pmatrix} -j1 & 0 & 0 \\ 0 & -j0.8 & 0 \\ 0 & 0 & -j0.5 \\ -j2.5 & j2.5 & 0 \\ 0 & j5 & -j5 \\ j5 & 0 & -j5 \end{pmatrix}$$

$$Y_{\text{bus}} = A^T YA = \begin{pmatrix} -j8.5 & j2.5 & j5 \\ j2.5 & -j8.3 & j5 \\ j5 & j5 & -j10.5 \end{pmatrix}$$

The advantages of using bus admittance matrix for power flow studies are as follows:

- Data preparation is simple.
- Bus admittance matrix can be easily formed and modified for network changes such as addition or deletion of lines, transformers, etc.

- Unlike bus impedance matrix which is a full matrix, bus admittance matrix is highly sparse and therefore reduces computer memory and execution time.

For an n-bus system, the number of distinct elements in the full matrix is n^2, while due to symmetry the maximum number of distinct elements in Y_{bus} matrix is $n(n + 1)/2$, the total of n natural numbers. The number of distinct elements is 1 in row 1, 2 in row 2, etc. and n in last row, triangular matrix elements.

3.5 Power Flow Solution

Power flow solution is the solution of the network under steady state conditions (load and generation are fixed at a particular time snap shot) subject to certain inequality constraints under which the system operates. These constraints can be in the form of bus voltages, line loading, etc.

3.5.1 Assumptions in Power Flow Studies

The following assumptions are made in the power flow studies:

- The system is balanced. Per phase analysis is possible in the place of three-phase analysis.
- Variation in load, generation and other parameters is neglected.
- Mutual coupling between transmission lines is neglected.

3.5.2 Steps Involved in Power Flow Solution

The steps followed in power flow solution are:

Step 1   Representation of the system by one-line diagram.
Step 2   Determination of impedance diagram.
Step 3   Formulation of power flow equations.
Step 4   Solution of power flow equations.

In the discussions, voltages at buses, i.e. at the transmission line ends are considered.

3.6 Power Flow Equation

The power injected at any bus p shown in Figure 3.6 is given by

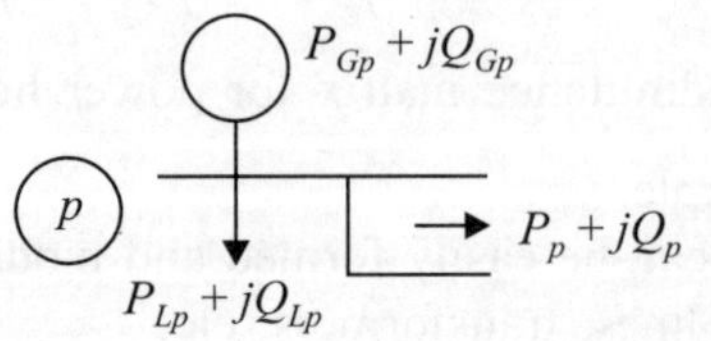

Figure 3.6 Complex power balance.

$$S_p = S_{Gp} - S_{Lp}$$
$$= P_p + jQ_p = V_p I_p^* \tag{3.13}$$

(as lagging VAR is assumed positive)
where

$$S_{Gp} = \text{Power Generation at bus } p$$
$$S_{Lp} = \text{Load at bus } p$$
$$S_p^* = V_p^* \, I_p \tag{3.14}$$

Using Eq. (3.2),

$$P_p - jQ_p = V_p^* \sum_{q=1}^{n} Y_{pq} V_q \qquad p = 1, 2, ..., n \tag{3.15}$$

Equation (3.15) is the power flow equation.

3.7   Computation of Slack Bus Power, Line Flows and Losses

3.7.1   Computation of Slack Bus Power

The unknowns at the slack bus viz., P and Q can be computed from the voltage known from the power flow studies.

$$S_s^* = P_s - jQ_s = V_s^* \sum_{q=1}^{n} Y_{sq} V_q$$

3.7.2   Computation of Line Flows

Consider the line connecting buses p and q as shown in Figure 3.7. The line is represented by equivalent series admittance $y_{pq}(\text{se})$ and line charging shunt admittances $y_{pq}(\text{sh})$ equally distributed at both ends of the line.

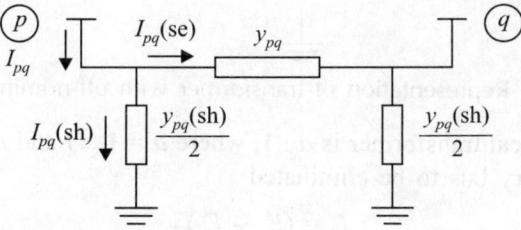

Figure 3.7   Line connecting two buses.

$$I_{pq} = I_{pq}(\text{se}) + I_{pq}(\text{sh})$$

$$= (V_p - V_q) y_{pq} + V_p \frac{y_{pq}(\text{sh})}{2}$$

$$S_{pq}^* = P_{pq} - jQ_{pq} = V_p^* I_{pq}$$

i.e.
$$S_{pq}^* = V_p^* \left[(V_p - V_q) y_{pq} + V_p \frac{y_{pq}(\text{sh})}{2} \right] \tag{3.16}$$

3.7.3 Computation of Transmission Line Loss

In any transmission line

Sending end power = Receiving end power + Loss
$$S_{pq} = -S_{qp} + S_{pq \text{ loss}}$$
∴ pq line loss = $S_{pq} + S_{qp}$ $\tag{3.17}$

3.7.4 Computation of Total Loss

Generation meets the load after incurring losses.

∴
$$\text{Total loss} = \sum_{p=1}^{n} S_{Gp} - \sum_{p=1}^{n} S_{Lp} \tag{3.18}$$

It may be noted that total loss is the sum of transmission losses and shunt losses.

3.8 Representation of Transformer with Off-nominal Turns Ratio

Transformer with off-nominal turns ratio can be represented by its leakage admittance in series with an ideal auto-transformer as shown in Figure 3.8.

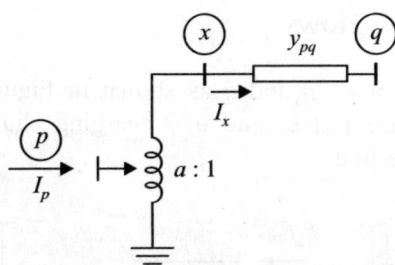

Figure 3.8 Representation of transformer with off-nominal turns ratio.

The turns ratio of ideal transformer is $a : 1$, where $a = 1 \pm t$ and t is the per-unit tap setting. Let x be the auxiliary bus to be eliminated.

$$I_x = (E_x - E_q) y_{pq}$$

Since the auto-transformer is ideal, there is no power loss in the transformer.

$$E_p^* I_p = E_x^* I_x$$

$$I_p = \left(\frac{E_x}{E_p} \right)^* I_x = \left(\frac{1}{a} \right)^* I_x = \left(\frac{1}{a} \right) I_x \quad \text{since } a \text{ is real}$$

$$= \left(\frac{1}{a} \right)(E_x - E_q) y_{pq} = \left(\frac{1}{a} \right)\left(\frac{E_p}{a} - E_q \right) y_{pq}$$

$$= \frac{y_{pq}}{a^2} E_p - \frac{y_{pq}}{a} E_q$$

In general

$$I_p = Y_{pp} E_p + Y_{pq} E_q$$

$$\therefore \qquad Y_{pp} = \frac{y_{pq}}{a^2} \quad \text{and} \quad Y_{pq} = -\frac{y_{pq}}{a}$$

Similarly

$$I_q = (E_q - E_x) y_{qp}$$

$$= \left(E_q - \frac{E_p}{a} \right) y_{pq}$$

$$= y_{pq} E_q - \frac{y_{pq}}{a} E_p$$

$$\Rightarrow \qquad Y_{qq} = y_{pq} \quad \text{and} \quad Y_{qp} = -\frac{y_{pq}}{a}$$

Since $y_{pq} = y_{qp}$ we can represent a π-equivalent of the above transformer which is shown in Figure 3.9.

$$Y_{pq} = Y_{qp} = -\frac{y_{pq}}{a}$$

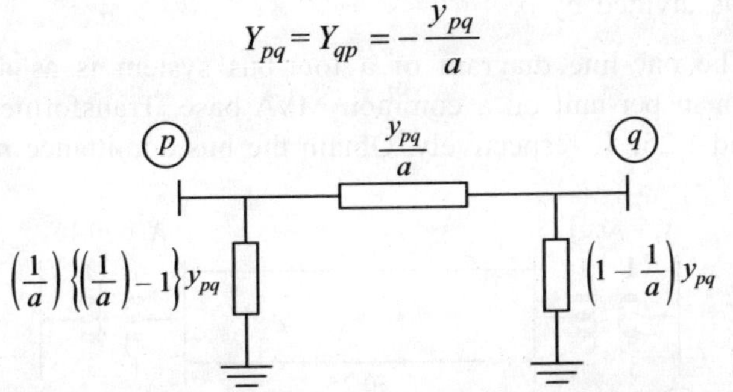

Figure 3.9 π-equivalent of transformer.

Hence series element is

$$-\left(-\frac{y_{pq}}{a} \right) = \frac{y_{pq}}{a}$$

Shunt elements Y_{pp} and Y_{qq} are calculated as follows:

$$Y_{pp} = \frac{y_{pq}}{a^2} = y_{pp} + \frac{y_{pq}}{a}$$

$$y_{pp} = \left(\frac{1}{a} \right) \left\{ \left(\frac{1}{a} \right) - 1 \right\} y_{pq}$$

$$Y_{qq} = y_{pq} = y_{qq} + \frac{y_{pq}}{a}$$

$$y_{qq} = \left\{ 1 - \left(\frac{1}{a} \right) \right\} y_{pq}$$

If in a system, this transformer is connected between buses p and q with tap on p side,

$$Y_{pp} = y_{p1} + y_{p2} + \cdots + \frac{y_{pq}}{a} + \left(\frac{1}{a} \right) \left\{ \left(\frac{1}{a} \right) - 1 \right\} y_{pq} + \cdots + y_{pn}$$

$$= y_{p1} + y_{p2} + \cdots + \frac{y_{pq}}{a^2} + \cdots + y_{pn} \tag{3.19}$$

$$Y_{pq} = - \frac{y_{pq}}{a} \tag{3.20}$$

$$Y_{qq} = y_{q1} + y_{q2} + \cdots + \frac{y_{pq}}{a} + \left\{ 1 - \left(\frac{1}{a} \right) \right\} y_{pq} + \cdots + y_{qn}$$

$$= y_{q1} + y_{q2} + \cdots + y_{pq} + \cdots + y_{qn} \tag{3.21}$$

That is, for computing self admittance, on the tapped side the transformer admittance value is divided by a^2 and on the untapped side it remains as it is. For the transfer admittance, the admittance value gets divided by a.

EXAMPLE 3.2 The one-line diagram of a four-bus system is as shown in Figure 3.10. Reactances are given in per-unit on a common MVA base. Transformers T_1 and T_2 have tap settings of 0.8: 1 and 1.25: 1, respectively. Obtain the bus admittance matrix.

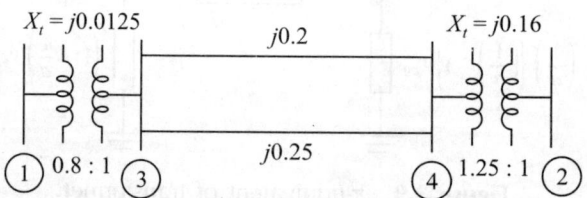

Figure 3.10 One-line diagram for Example 3.2.

Solution

$$y_{T_1} = \frac{1}{j0.0125} = - j80$$

$$y_{T_2} = \frac{1}{j0.16} = - j6.25$$

$$y_{L_1} = \frac{1}{j0.2} = - j5$$

$$y_{L_2} = \frac{1}{j0.25} = - j4$$

$$Y_{11} = \frac{y_{T_1}}{a_{T_1}^2} = -\left(\frac{-j80}{0.8^2}\right) = -j125$$

$$Y_{12} = 0 = Y_{21}$$

$$Y_{13} = \frac{-y_{T_1}}{a_{T_1}} = -\left(\frac{-j80}{0.8}\right) = j100 = Y_{31}$$

$$Y_{14} = 0 = Y_{41}$$

$$Y_{22} = y_{T_2} = -j6.25$$

$$Y_{23} = 0 = Y_{32}$$

$$Y_{24} = \frac{-y_{T_2}}{a_{T_2}} = -\left(\frac{-j6.25}{1.25}\right) = j5 = Y_{42}$$

$$Y_{33} = y_{T_1} + y_{L_1} + y_{L_2} = -j80 - j5 - j4 = -j89$$

$$Y_{34} = -(y_{L_1} + y_{L_2}) = -(-j5 - j4) = j9 = Y_{43}$$

$$Y_{44} = \frac{y_{T_2}}{a_{T_2}^2 + y_{L_1} + y_{L_2}} = \left(\frac{-j6.25}{1.25^2 - j5 - j4}\right) = -j13$$

$$Y_{bus} = \begin{pmatrix} -j125 & 0 & j100 & 0 \\ 0 & j6.25 & 0 & j5 \\ j100 & 0 & j89 & j9 \\ 0 & 5 & 9 & -j13 \end{pmatrix}$$

Short Answer Questions

1. **What is a bus?**

   The meeting point of various components in a power system is called a *bus*. The bus is a conductor having negligible impedance. The bus is considered as point of constant voltage in a power system.

   From network point of view, when one of the nodes of a power network is taken as the reference, the other nodes of the network are called *buses*.

2. **What is bus admittance matrix?**

   The matrix consisting of the self and mutual admittances of the network of a power system is called *bus admittance matrix*. It is formed from the nodal equations of the power system and is symmetrical.

3. **Name the diagonal and off-diagonal elements of bus admittance matrix.**

   The diagonal elements of a bus admittance matrix are called *self admittances* of the buses and obtained by adding all the admittances connected to that bus. The off-diagonal elements are called *mutual* or *transfer admittances* of the buses and obtained by negating the admittance between the two buses under consideration.

4. What is power flow study?

   Solving the power system network to obtain the voltages (magnitude and phase angle) at all the buses, power (real and reactive) flowing over all the lines and transformers, losses and power injected at all the buses is called *power flow study*.

5. What is the information that is obtained from a power flow study?

   The voltage (magnitude and phase angle) at all the buses, power (real and reactive) flowing over all the lines and transformers, losses and power injected at all the buses is obtained from power flow study.

6. What is the need for power flow study?

   The power flow study of a power system is required to decide optimum operation of existing system with adequate security and for planning the future expansion of the system with sufficient reliability. It is also essential for designing a new power system. The solution obtained from the power flow study also gives the initial conditions of the system for carrying out fault analysis and stability studies to evaluate the transient behaviour of the system.

7. What are the steps involved in power flow study?

   The sequence of steps involved in power flow study is:

   Step 1   Representation of the system by one-line diagram.

   Step 2   Determining the impedance diagram.

   Step 3   Formulation of non-linear algebraic power flow equations.

   Step 4   Solution of power flow equations by iterative method.

8. What are the parameters associated with each bus in a system?

   Each bus in a system is associated with four parameters namely magnitude and phase angle of voltage and real and reactive power.

9. What are the different types of buses in a power system?

   In a power flow problem, two out of the four quantities—magnitude and phase angle of voltage, and real and reactive power are specified and the remaining two are determined by the solution of power flow equations. Accordingly, the buses are classified into three types, namely load bus, generation or voltage controlled bus and slack or swing or reference bus. The quantities specified and to be determined for the different types of buses are given in Table 3.1.

10. What is the need for slack bus and how is it selected?

    In a power system generation and load are known, *apriori*, but transmission and transformation losses will be known only after the power flow solution is complete because losses depend on the relative location and dispatch of generation with respect to load. It is therefore required to consider one of the generation buses as slack bus which will supply the power losses (active and reactive).

    The slack bus is selected based on the following considerations:

    • Maximum power

    • Location near centre of gravity of load

    • Costly generation

11. What is P–Q bus in power flow analysis?
    A bus is called P–Q bus when real and reactive components of power are specified for the bus and the voltage is allowed to vary within permissible limits.

12. What do you mean by flat voltage start?
    In iterative methods of power flow solution, the initial voltages of load buses are assumed as $1 + j0$ pu. This is referred to as flat voltage start.

13. Define primitive network.
    Each source and the shunt admittance connected across it are represented by a single element, called *primitive network*.

14. What is an incidence matrix?
    An incidence matrix represents, in general, interconnection of the elements with respect to the nodes of a network.

15. What is the limitation of direct inspection method for forming Y_{bus}?
    If there is mutual coupling between the elements of a power network, direct inspection method cannot be applied for forming Y_{bus}.

16. How can we find the order of the bus incidence matrix?
    Number of branches and buses give the order of the bus incidence matrix.

17. What is off-nominal turns ratio?
    The turns ratio of a regulating transformer used to maintain the voltage of a bus is called *off-nominal* turns ratio. This is different from the ratio used to select the base voltages on the two sides of the transformer.

18. Write the power flow equation.
    The power flow equation is

    $$P_p - jQ_p = V_p^* \sum_{q=1}^{n} Y_{pq} V_q \qquad p = 1, 2, ..., n$$

    where,
    P_p = real power injection at bus p
    Q_p = reactive power injection at bus p
    Y_{pq} = element (p, q) of bus admittance matrix
    V_q = voltage phasor at bus q

19. How is the off-nominal turns ratio transformer represented in bus admittance matrix?
    For computing self admittance, on the tapped side the transformer admittance value is divided by a^2 and on the untapped side it remains as it is. For the transfer admittance, the admittance value gets divided by a.

Exercises

3.1 A simple power system is shown in Figure E3.1, where impedances are expressed in per-unit on a common base and resistances are neglected. Convert network impedances to admittances and obtain the bus admittance matrix by inspection.

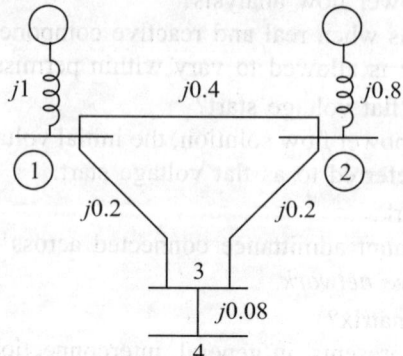

Figure E3.1   One-line diagram for Problem 3.1.

$$\text{Ans.} \quad \begin{bmatrix} -j8.5 & j2.5 & j5 & 0 \\ j2.5 & j8.75 & j5 & 0 \\ j5 & j5 & j22.5 & j12.5 \\ 0 & 0 & j12.5 & -j12.5 \end{bmatrix}$$

3.2 A power system network is shown in Figure E3.2. The generators at buses 1 and 2 are represented by their equivalent current sources with their reactances expressed in per-unit on a 100 MVA base. The lines are represented by π-model, where series reactances and shunt reactances are also expressed in per-unit on a 100 MVA base. The loads at buses 3 and 4 are expressed in MW and MVAR respectively. Assuming a voltage magnitude of 1.0 per-unit at buses 3 and 4, convert the loads to per unit impedances. Convert network impedances to admittances and obtain the bus admittance matrix.

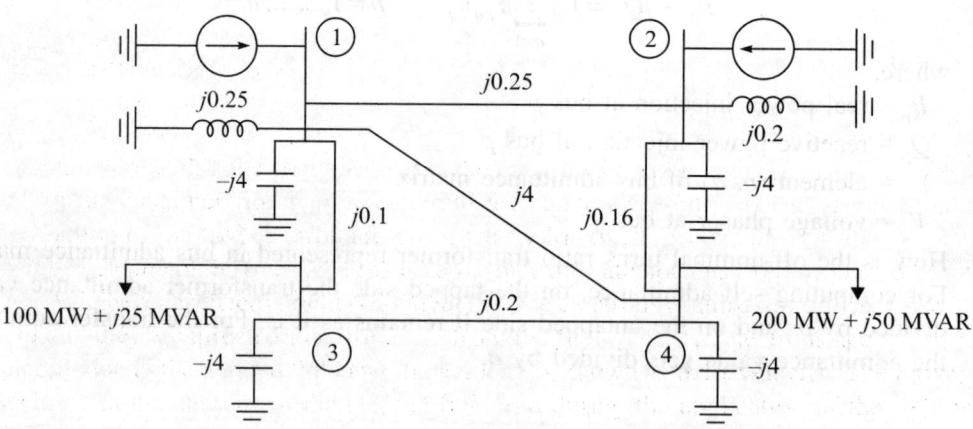

Figure E3.2   One-line diagram for Problem 3.2.

$$\text{Ans.} \quad \begin{bmatrix} 0-j20.25 & 0+j4 & 0+j10 & 0+j2.5 \\ 0+j4 & 0-j15 & 0 & 0+j6.25 \\ 0+j10 & 0 & 1-j15 & j5 \\ 0+j2.5 & 0+j6.25 & j5 & 2-j14 \end{bmatrix}$$

3.3 A power system network is shown in Problem E3.3.

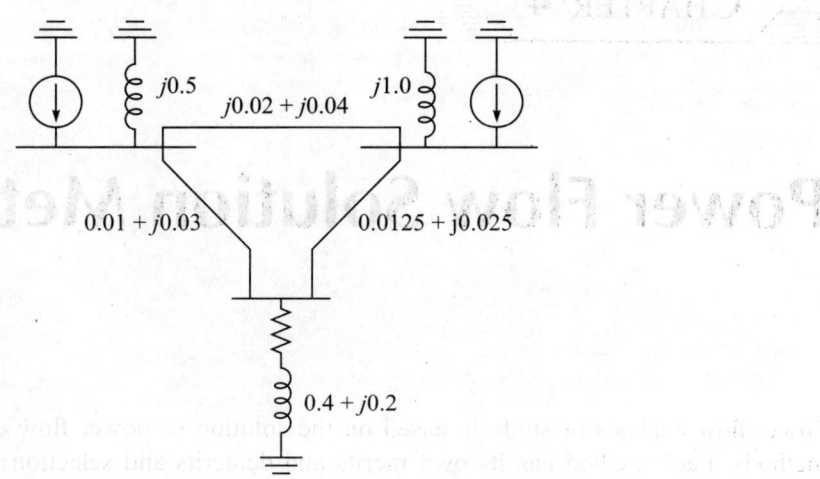

Figure E3.3 One-line diagram for Problem 3.3.

The values marked are impedances in per-unit on a base of 100 MVA. Determine the bus admittance matrix by inspection method.

CHAPTER 4

Power Flow Solution Methods

Power flow analysis or study is based on the solution of power flow equations using various methods. Each method has its own merits and demerits and selection of the method depends on the area of application and the conditions of the system.

4.1 Power Flow Solution Methods

Power flow equations are nonlinear algebraic equations and can be solved only by iterative methods. The following are the iterative methods used to solve these equations:

- Gauss method
- Gauss–Seidal method
- Newton–Raphson method (polar co-ordinates)
- Newton–Raphson method (rectangular coordinates)
- Decoupled method
- Fast decoupled power flow method

4.1.1 Gauss and Gauss–Seidal Methods

The power flow equation given by Eq. (3.15) may be rewritten as

$$\sum_{q=1}^{n} Y_{pq} V_q = \frac{P_p - jQ_p}{V_p^*}$$

$$Y_{pp} V_p + \sum_{\substack{q=1 \\ \neq p}}^{n} Y_{pq} V_q = \frac{P_p - jQ_p}{V_p^*}$$

$$Y_{pp} V_p = \frac{P_p - jQ_p}{V_p^*} - \sum_{\substack{q=1 \\ \neq p}}^{n} Y_{pq} V_q$$

$$V_p = \frac{1}{Y_{pp}} \left(\frac{P_p - jQ_p}{V_p^*} - \sum_{\substack{q=1 \\ \neq p}}^{n} Y_{pq} V_q \right) \tag{4.1}$$

The voltage at bus p during iteration $(k + 1)$ is therefore

$$V_p^{k+1} = \frac{1}{Y_{pp}} \left[\frac{P_p - jQ_p}{\left(V_p^k\right)^*} - \sum_{\substack{q=1 \\ \neq p}}^{n} Y_{pq} V_q^k \right] \tag{4.2}$$

Equation (4.2) is used in **Gauss method**. Same set of voltage values is used throughout a complete iteration. The criterion to stop iteration, that is, to reach convergence is when the maximum of the magnitude differences between the voltages of consecutive iterations becomes less than a certain specified tolerance or mismatch, say $|\Delta V|_{max} < \varepsilon = 0.001$.

As the bus voltages V_q up to $(p - 1)$ have already been computed during the iteration $(k + 1)$, using these values in stead of the values computed at the end of kth iteration can reduce the number of iterations resulting in quicker convergence.

Rewriting Eq. (4.1)

$$V_p = \frac{1}{Y_{pp}} \left(\frac{P_p - jQ_p}{\left(V_p\right)^*} - \sum_{q=1}^{p-1} Y_{pq} V_q - \sum_{q=p+1}^{n} Y_{pq} V_q \right) \tag{4.3}$$

Bus voltage V_p during $(k + 1)$th iteration is

$$V_p^{k+1} = \frac{1}{Y_{pp}} \left(\frac{P_p - jQ_p}{\left(V_p^k\right)^*} - \sum_{q=1}^{p-1} Y_{pq} V_q^{k+1} - \sum_{q=p+1}^{n} Y_{pq} V_q^k \right) \tag{4.4}$$

Equation (4.4) is used in **Gauss–Seidal method**. Each new voltage value obtained is immediately substituted to calculate the voltage of next bus.

The number of iterations is still large and the process of convergence to obtain the solution is slow. Convergence can be speeded up by using the acceleration factor and the voltage during consecutive iteration is modified as

$$V_{p\,acc}^{k+1} = V_p^k + \alpha(V_p^{k+1} - V_p^k) \tag{4.5}$$

where α is the acceleration factor which is real and greater than 1.

Trial power flow studies give suitable value of α. A wrong choice of α may result in either slower convergence or even divergence. A generally recommended value for most of the systems lies between 1.4 and 1.6.

To reduce the computation time the following arithmetic operations, which do not change with iterations can be performed in advance for the load buses:

$$A_p = \frac{P_p - jQ_p}{Y_{pp}}, \qquad B_p = \frac{Y_{pq}}{Y_{pp}}$$

However, for PV buses these are to be calculated during the iterative process whenever the PV buses become load buses due to Q-limit violation.

At PV buses the magnitude of the voltage is to be maintained at specified value, while the phase angle will be as per Eq. (4.5).

The *PV* buses need adequate reactive power support to maintain the voltage magnitude at the specified value. If the Q value violates the reactive power limits, then during that iteration the *PV* bus will be treated as *PQ* bus. Q_p is calculated from the following equation as derived from Eq. (3.15)

$$Q_p = -\operatorname{Im}\left[V_p^* \sum_{q=1}^{n} Y_{pq} V_q \right]$$

4.2 Power Flow Solution using Gauss–Seidal Method

4.2.1 Algorithm/Steps for Gauss–Seidal Method

Step 1 Read system data, tolerance for mismatch ε and acceleration factor α.

Step 2 Formulate Y_{bus}.

Step 3 Select the slack bus s, if not specified.

Step 4 Assume initial bus voltages for all buses.

$$V_p^0 = V_{p\ spec} \angle 0° \text{ at all generation buses.}$$
$$V_p^0 = 1 \angle 0° \text{ at all load buses (flat start).}$$

Step 5 Set iteration count $k = 0$.

$$P_p = P_{Gp} - P_{Lp}; \quad Q_p = Q_{Gp} - Q_{Lp};$$
$$Q_{p\ min} = Q_{Gp\ min} - Q_{Lp}; \quad Q_{p\ max} = Q_{Gp\ max} - Q_{Lp}.$$

Step 6 Set bus count $p = 1$.

Step 7 If p is the slack bus, go to Step 13.

Step 8 If p is load bus, go to Step 10.

Step 9 Compute $Q_p = -\operatorname{Im}\left[V_p^* \left\{ \sum_{q=1}^{p-1} Y_{pq} V_q^{k+1} + \sum_{q=p}^{n} Y_{pq} V_q^k \right\} \right]$

Check for Q limit violation [see Figure 4.1]

if $Q_{p\ min} < Q_p < Q_{p\ max}$, then $Q_p = Q_{p\ cal}$.

if $Q_p < Q_{p\ min}$, then $Q_p = Q_{p\ min}$

if $Q_p > Q_{p\ max}$, then $Q_p = Q_{p\ max}$

When Q limit is violated, then the bus is treated as *PQ* bus during this iteration; $V_p^k = 1\angle 0°$.

Step 10 Compute V_p^{k+1} from Eq. (4.4), which is reproduced below.

$$V_p^{k+1} = \frac{1}{Y_{pp}}\left(\frac{P_p - jQ_p}{(V_p^k)^*} - \sum_{q=1}^{p-1} Y_{pq} V_q^{k+1} - \sum_{q=p+1}^{n} Y_{pq} V_q^k \right)$$

Step 11 Compute $\Delta V_p^k = V_p^{k+1} - V_p^k$. Calculate the accelerated bus voltages of load and Q limit violated *PV* buses.

$$V_{p\ acc}^{k+1} = V_p^k + \alpha \Delta V_p^k$$

Step 12   For PV bus having Q_p within limits, $|V_{p\,acc}^{k+1}| = |V_{p\,spec}|$.
Step 13   If p is less than or equal to number of buses, increment the bus count and go to Step 7.
Step 14   Check for convergence. If $|\Delta V|_{max} > \varepsilon$, increment iteration count and go to Step 6.
Step 15   Calculate slack bus power, reactive power at PV buses, line flows and losses.

Flow chart for Gauss–Seidal Method is shown in Figure 4.1.

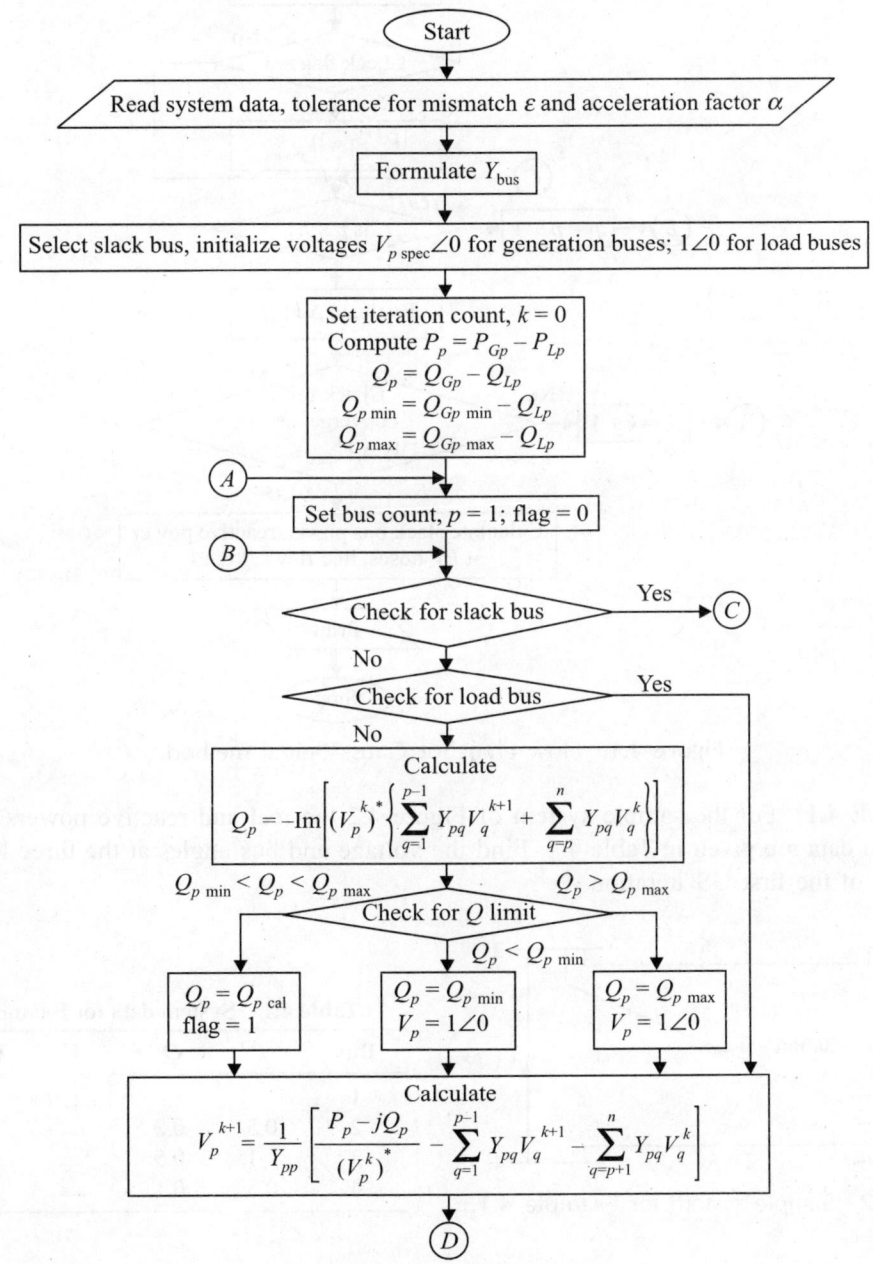

Figure 4.1   (Contd.)

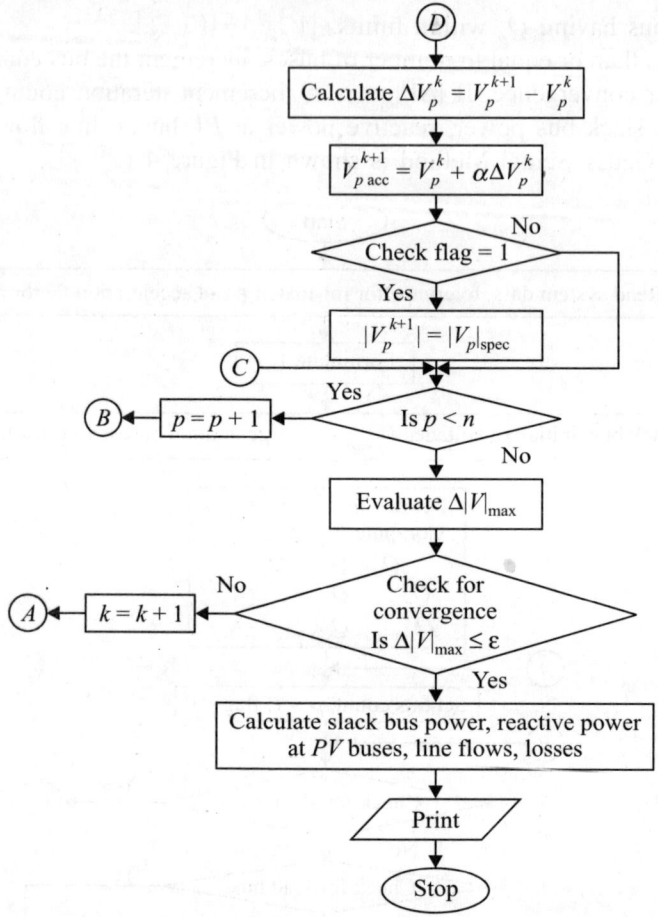

Figure 4.1   Flow chart for Gauss–Seidal method.

EXAMPLE 4.1   For the sample system of Figure 4.2 the real and reactive powers are listed and system data are given in Table 4.1. Find the voltage and bus angles at the three load buses at the end of the first GS iteration.

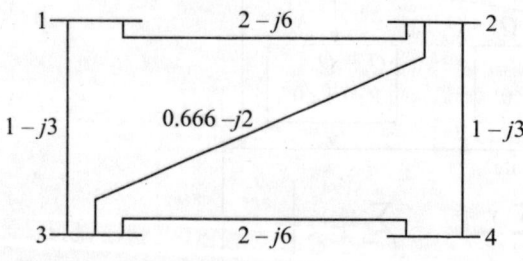

Figure 4.2   Sample system for Example 4.1.

Table 4.1   System data for Example 4.1

Bus	P	Q	V	Remarks
1	–	–	1.04∠0°	Slack
2	0.5	–0.2	—	PQ
3	–1	0.5	—	PQ
4	0.3	–0.1	—	PQ

Solution From the impedances of the lines, admittances are calculated.

$$Y_{bus} = \begin{bmatrix} Y_{11} & Y_{12} & Y_{13} & Y_{14} \\ Y_{21} & Y_{22} & Y_{23} & Y_{24} \\ Y_{31} & Y_{32} & Y_{33} & Y_{34} \\ Y_{41} & Y_{42} & Y_{43} & Y_{44} \end{bmatrix} = \begin{bmatrix} y_{12} + y_{13} & -y_{12} & -y_{13} & 0 \\ -y_{21} & y_{21} + y_{23} + y_{24} & -y_{23} & -y_{24} \\ -y_{31} & -y_{32} & y_{31} + y_{32} + y_{34} & -y_{34} \\ 0 & -y_{42} & -y_{43} & y_{42} + y_{43} \end{bmatrix}$$

$$= \begin{bmatrix} 3 - j9 & -2 + j6 & -1 + j3 & 0 \\ -2 + j6 & 3.666 - j11 & -0.666 + j2 & -1 + j3 \\ -1 + j3 & -0.666 + j2 & 3.666 - j11 & -2 + j6 \\ 0 & -1 + j3 & -2 + j6 & 3 - j9 \end{bmatrix}$$

Initialise bus voltages:

$$V_1 = 1.04\angle 0°$$

$$V_2^0 = V_3^0 = V_4^0 = 1\angle 0° \text{ (Flat start)}$$

Bus voltage at the end of first iteration are calculated from

$$V_p^{k+1} = \frac{1}{Y_{pp}}\left(\frac{P_p - jQ_p}{(V_p^k)^*} - \sum_{q=1}^{p-1} Y_{pq} V_q^{k+1} - \sum_{q=p+1}^{n} Y_{pq} V_q^k\right)$$

$$V_2^1 = \frac{1}{Y_{22}}\left[\frac{P_2 - jQ_2}{(V_2^0)^*} - Y_{21}V_1 - Y_{23}V_3^0 - Y_{24}V_4^0\right]$$

$$= \frac{1}{3.666 - j11}\left[\frac{0.5 + j0.2}{1\angle -0°} - 1.04(-2 + j6) - (-0.666 + j2) - (-1 + j3)\right]$$

$$= \frac{4.246 - j11.04}{3.666 - j11} = 1.019 + j0.046 = 1.02\angle 2.58°$$

$$V_3^1 = \frac{1}{Y_{33}}\left[\frac{P_3 - jQ_3}{(V_3^0)^*} - Y_{31}V_1 - Y_{32}V_2^1 - Y_{34}V_4^0\right]$$

$$= \frac{1}{3.666 - j11}\left[\frac{-1 - j0.5}{1\angle -0°} - 1.04(-1 + j3) - (0.666 + j2) \times 1.02\angle 2.58° - (-2 + j6)\right]$$

$$= \frac{2.81 - j11.627}{3.666 - j11} = 1.028 + j0.087 = 1.0317\angle 4.84°$$

$$V_4^1 = \frac{1}{Y_{44}}\left[\frac{P_4 - jQ_4}{(V_4^0)^*} - Y_{41}V_1 - Y_{42}V_2^1 - Y_{43}V_3^1\right]$$

$$= \frac{1}{3 - j9}\left[\frac{0.3 + j0.1}{1\angle -0°} - (-1 + j3)1.02\angle 2.2° - (-2 + j6)1.0317\angle 4.84°\right]$$

$$= \frac{2.991 - j9.253}{3 - j9} = 1.025 + j0.0093 = 1.025\angle 0.52°$$

EXAMPLE 4.2 For the system shown in Figure 4.3 and data given in Table 4.2, determine the voltage at the end of the first iteration by Gauss–Seidal method and also find the slack bus power, line flows and transmission loss. Assume MVA base as 100.

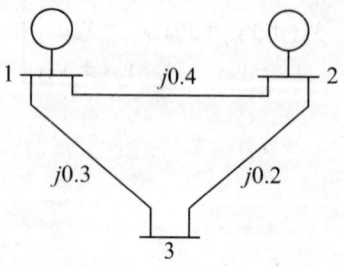

Figure 4.3 System for Example 4.2.

Table 4.2 System data for Example 4.2

Bus	Voltage	Generator		Load		Q_{min}	Q_{max}
		P	Q	P	Q		
1	$1.05\angle 0$	—	—	—	—	—	—
2	1.02	0.3	—	—	—	−10	100
3	—	—	—	0.4	0.2	—	—

Solution

Step 1 Form Y_{bus}

$$Y_{bus} = \begin{bmatrix} \dfrac{1}{j0.4} + \dfrac{1}{j0.3} & -\dfrac{1}{j0.4} & -\dfrac{1}{j0.3} \\[2mm] -\dfrac{1}{j0.4} & \dfrac{1}{j0.4} + \dfrac{1}{j0.2} & -\dfrac{1}{j0.2} \\[2mm] -\dfrac{1}{j0.3} & -\dfrac{1}{j0.2} & \dfrac{1}{j0.3} + \dfrac{1}{j0.2} \end{bmatrix}$$

$$= \begin{bmatrix} -j5.8333 & j2.5 & j3.3333 \\ j2.5 & -j7.5 & j5 \\ j3.3333 & j5 & -j8.3333 \end{bmatrix}$$

Step 2 Initialise bus voltages

$$V_1 = 1.05\angle 0$$
$$V_2^0 = 1.02\angle 0$$
$$V_3^0 = 1\angle 0 \text{ (flat start)}$$
$$P_2 = P_{G2} - P_{L2} = 0.3 - 0 = 0.3 \text{ pu}$$
$$P_3 = P_{G3} - P_{L3} = 0 - 0.4 = -0.4 \text{ pu}$$
$$Q_3 = Q_{G3} - Q_{L3} = 0 - 0.2 = -0.2 \text{ pu}$$

Step 3 Calculate Q for generating buses

$$Q_p = -\text{Im}\left[(V_p^k)^* \left\{ \sum_{q=1}^{p-1} Y_{pq} V_q^{k+1} + \sum_{q=p}^{n} Y_{pq} V_q^k \right\} \right]$$

$$Q_p^0 = -\text{Im}[(V_2^0)^* \{ Y_{21}V_1 + Y_{22}V_2^0 + Y_{23}V_3^0 \}]$$

$$= -\text{Im}[1.02\angle 0°\{ j2.5 \times 1.05\angle 0° + (-j7.5)1.02\angle 0° + j5 \times 1\angle 0° \}]$$

$$= -\text{Im}[1.02\angle 0°(+j2.625 - j7.65 + j5)] = 0.025 \text{ pu}$$

$$Q_{2\,min} < Q_2 < Q_{2\,max}$$

Thus, Q_2 is within the limits and bus 2 acts as generating bus.

Step 4   Calculate voltages at the next iteration

$$V_p^{k+1} = \frac{1}{Y_{pp}}\left[\frac{P_p - jQ_p}{(V_p^k)^*} - \sum_{q=1}^{p-1} Y_{pq}V_q^{k+1} - \sum_{q=p+1}^{n} Y_{pq}V_q^k\right]$$

$$V_2^1 = \frac{1}{Y_{22}}\left[\frac{P_2 - jQ_2}{(V_2^0)^*} - Y_{21}V_1 - Y_{23}V_3^0\right]$$

$P_2 = 0.3$ pu (given); $Q_2 = 0.025$ pu (calculated)

$$V_2^1 = \frac{1}{-j7.5}\left[\frac{0.3 - j0.025}{1.02\angle -0°} - j2.5 \times 1.05\angle 0° - j5 \times 1\angle 0°\right]$$

$$= 1.0199 + j0.0392$$

$$= 1.0207\angle 2.2°$$

Since bus is *PV* bus having *Q* within limits, magnitude of V_2 will be maintained constant at the specified value. Hence

$$V_2^1 = V_{2\,spec}\angle \delta_{2\,cal} = 1.02\angle 2.2° = 1.0192 + j0.0392$$

$$V_3^1 = \frac{1}{Y_{33}}\left[\frac{P_3 - jQ_3}{(V_3^0)^*} - Y_{31}V_1 - Y_{32}V_2^1\right]$$

$$= \frac{1}{-j8.3333}\left[\frac{-0.4 + j0.2}{1\angle -0°} - j3.3333 \times 1.05\angle 0° - j5 \times 1.02\angle 2.2°\right]$$

$$= \frac{1}{-j8.3333}[-0.4 + j0.2 - j3.4999 - j5.096 + 0.196]$$

$$= 1.0075 - j0.0244$$

$$= 1.0078\angle -1.39°$$

Step 5   Slack bus power

$$S_s^* = P_s - jQ_s = V_s^* \sum_{q=1}^{n} Y_{sq}V_q$$

$$S_1^* = V_1^*[Y_{11}V_1 + Y_{12}V_2^1 + Y_{13}V_3^1]$$

$$= 1.05\angle 0°[-j5.8333 \times 1.05\angle 0° + j2.5(1.0192 + j0.0392)$$

$$+ j3.3333(1.0075 - j0.0244)]$$

$$= -0.0175 - j0.2295$$

$$P_1 = -0.0175\text{ pu} = -1.75\text{ MW}$$

$$Q_1 = 0.2295\text{ pu} = 22.95\text{ MVAR}$$

Step 6   Calculate the line flows as given in Table 4.3.

Table 4.3 Line flows for Example 4.2

Bus no. From	To	$S_{pq}^* = P_{pq} - jQ_{pq} = V_p^*(V_p - V_q)y_{pq} + \dfrac{V_p^2 y_{pq(sh)}}{2}$ $= y_{pq}(V_p^2 - V_p^* V_q)$ if $y_{pq(sh)} = 0$
1	2	$j2.5(1.05^2 - 1.05\angle{-0°} \times 1.02\angle2.2°) = -0.1029 - j0.0808$ pu $= -10.29$ MW $- 8.08$ MVAR
2	1	$j2.5(1.02^2 - 1.02\angle{-2.2°} \times 1.05\angle0°) = 0.1029 + j0.0746$ pu $= 10.29$ MW $+ 7.46$ MVAR
2	3	$j5(1.02^2 - 1.02\angle{-2.2°} \times 1.0078\angle{-1.39°}) = 0.3218 - j0.072$ pu $= 32.18$ MW $- 7.2$ MVAR
3	2	$j5(1.0078^2 - 1.0078\angle1.39° \times 1.02 \times 2.2°) = -0.3218 + j0.0512$ pu $= -32.18$ MW $+ 5.12$ MVAR
1	3	$j3.3333(1.05^2 - 1.05\angle{-0°} \times 1.0078\angle{-1.39°}) = 0.085 - j0.148$ pu $= 8.5$ MW $- 14.8$ MVAR
3	1	$j3.3333(1.0078^2 - 1.0078\angle1.39° \times 1.05\angle0°) = -0.085 + j0.1407$ pu $= -8.5$ MW $+ 14.07$ MVAR

Step 7   Transmission loss

For line p–q, $S_{pq\,loss} = S_{pq} + S_{qp}$

For line 1–2, $S_{12\,loss} = P_{12\,loss} + jQ_{12\,loss}$
$$= S_{12} + S_{21}$$
$$= -0.1029 + j0.0808 + 0.1029 - j0.0746$$
$$= 0 + j0.0061 = 0 \text{ MW} + 0.61 \text{ MVAR}$$

For line 2–3, $S_{23\,loss} = P_{23\,loss} + jQ_{23\,loss}$
$$= S_{23} + S_{32}$$
$$= 0.3218 + j0.072 - 0.3218 - j0.0512$$
$$= 0 + j0.0208 = 0 \text{ MW} + 2.08 \text{ MVAR}$$

For line 1–3, $S_{13\,loss} = P_{13\,loss} + jQ_{13\,loss}$
$$= S_{13} + S_{31}$$
$$= 0.085 + j0.148 - 0.085 - j0.1407$$
$$= 0 + j0.0073 = 0 \text{ MW} + 0.73 \text{ MVAR}$$

EXAMPLE 4.3 The reactive power constraint on generator bus –2 is changed to $10 \leq Q_2 \leq 100$ in Example 4.2. Determine the voltages at the end of the first iteration by GS method.

Solution First three steps (Steps 1–3) are same as in Example 4.2

$$Q_{2\,cal} = 0.025 \text{ pu} = 2.5 \text{ MVAR} < 10 \text{ MVAR} = Q_{2\,min}$$

∴   Bus –2 will be treated as PQ bus and $Q_2 = 10$ MVAR $= 0.1$ pu.

$$V_2^0 = 1\angle0°$$

Step 4

$$V_2^1 = \frac{1}{Y_{22}}\left[\frac{P_2 - jQ_2}{(V_2^0)^*} - Y_{21}V_1 - Y_{23}V_3^0\right]$$

$$= \frac{1}{-j7.5}\left[\frac{0.3 - j0.1}{1\angle{-0°}} - j2.5 \times 1.05\angle0° - j5 \times 1\angle0°\right]$$

$$= 1.03 + j0.04 = 1.0308\angle 2.22°$$

$$V_3^1 = \frac{1}{Y_{33}} \left[\frac{P_3 - jQ_3}{(V_3^0)^*} - Y_{31}V_1 - Y_{32}V_2^1 \right]$$

$$= \frac{1}{-j8.3333} \left[\frac{-0.4 + j0.2}{1\angle -0°} - j3.3333 \times 1.05\angle 0° - j5 \times 1.0308\angle 2.22° \right]$$

$$= 1.04 - j0.024 = 1.014\angle -1.36°$$

EXAMPLE 4.4   The reactive power constraint on generator bus −2 is changed to $0.004 < Q_2 < 0.01$ pu in Example 4.2. Determine the voltages at the end of first iteration by GS method.

Solution   First three steps are same as in Example 4.2

$$Q_{2\ cal} = 0.025 > 0.01 = Q_{2\ max}$$

∴   Bus −2 will be treated as PQ bus and $Q_2 = 0.01$ pu

$$V_2^0 = 1\angle 0°$$

Step 4

$$V_2^1 = \frac{1}{Y_{22}} \left[\frac{P_2 - jQ_2}{(V_2^0)^*} - Y_{21}V_1 - Y_{23}V_3^0 \right]$$

$$= \frac{1}{-j7.5} \left[\frac{0.3 - j0.01}{1\angle -0°} - j2.5 \times 1.05\angle 0° - j5 \times 1\angle 0° \right]$$

$$= 1.018 + j0.04 = 1.0187\angle 2.25°$$

$$V_3^1 = \frac{1}{Y_{33}} \left[\frac{P_3 - jQ_3}{(V_3^0)^*} - Y_{31}V_1 - Y_{32}V_2^1 \right]$$

$$= \frac{1}{-j8.3333} \left[\frac{-0.4 + j0.2}{1\angle -0°} - j3.3333 \times 1.05\angle 0° - j5 \times 1.0187\angle 2.25° \right]$$

$$= 1.0068 - j0.024 = 1.0071\angle -1.37°$$

4.3   Power Flow Solution using Newton–Raphson Method

The Gauss–Seidal method is very simple but the convergence becomes increasingly slow as the system size grows. The Newton–Raphson method gives solution that converges fast for both small and large systems, usually in less than 4 to 5 iterations although more number of calculations per iteration is required.

Newton–Raphson method using polar coordinates has many advantages over the use of rectangular coordinates.

1. The number of equations are less corresponding to the number of PV buses;
2. The dependence of P on δ and Q on $|V|$ is explicitly seen in polar form.

Newton–Raphson method is a successive approximation method based on Taylor series expansion.

4.3.1 Power Flow Model

Power injection at bus p is given by Eq. (3.14)

$$S_p^* = V_p^* I_p$$

$$P_p - jQ_p = V_p^* \sum_{q=1}^{n} Y_{pq} V_q \text{ using Eq. (3.2)}$$

$$= V_p \angle -\delta_p \sum_{q=1}^{n} Y_{pq} \angle \theta_{pq} V_q \angle \delta_q$$

$$= \sum_{q=1}^{n} V_p Y_{pq} V_q \angle (\theta_{pq} + \delta_q - \delta_p)$$

By equating real and imaginary parts, we have

$$P_p = \sum_{q=1}^{n} V_p Y_{pq} V_q \cos(\theta_{pq} + \delta_q - \delta_p) \qquad (4.6)$$

$$Q_p = -\sum_{q=1}^{n} V_p Y_{pq} V_q \sin(\theta_{pq} + \delta_q - \delta_p) \qquad (4.7)$$

Equations (4.6) and (4.7) constitute a set of nonlinear algebraic equations in terms of independent variables δ and $|V|$.

Power flow model in real variable form is compiled by following rules.

- For every bus whose phase angle δ is unknown, include respective real power balance equation [$(n-1)$ equations]
- For every bus whose voltage magnitude V is unknown, include respective reactive power balance equation that is, the reactive power balance equation is applicable only for load (PQ) buses. If m is number of PV buses, then the set of these equations is $(n - m - 1)$.

4.3.2 Power Flow Equations

Equations (4.6) and (4.7) are linearised by expanding in Taylor series around the starting values and neglecting higher order terms containing higher derivatives. We leave the calculations at slack bus, say bus number 1, since voltage magnitude and angle are specified at the slack bus. Next, since the voltage is constant ($\Delta V = 0$) at voltage controlled bus, the partial derivatives with respect to voltage magnitudes of these m buses are omitted. As the reactive power at the voltage controlled bus is not specified, ΔQ cannot be defined and so the partial derivatives of Q for the voltage controlled buses are also omitted.

$$P_p \cong P_p^0 + \left(\frac{\partial P_p}{\partial \delta_2}\right)^0 \Delta \delta_2^0 + \cdots + \left(\frac{\partial P_p}{\partial \delta_n}\right)^0 \Delta \delta_n^0 + \left(\frac{\partial P_p}{\partial |V_{m+2}|}\right)^0 \Delta |V_{m+2}|^0 + \cdots + \left(\frac{\partial P_p}{\partial |V_n|}\right)^0 \Delta |V_n|^0$$

$$\text{for } p = 1, 2, 3, ..., n$$

$$Q_p \cong Q_p^0 + \left(\frac{\partial Q_p}{\partial \delta_2}\right)^0 \Delta\delta_2^0 + \cdots + \left(\frac{\partial Q_p}{\partial \delta_n}\right)^0 \Delta\delta_n^0 + \left(\frac{\partial Q_p}{\partial |V_{m+2}|}\right)^0 \Delta|V_{m+2}|^0 + \cdots + \left(\frac{\partial Q_p}{\partial |V_n|}\right)^0 \Delta|V_n|^0$$

$$\text{for } p = m + 2, \ldots, n$$

where n is the total number of buses and m the number of voltage controlled buses.

$$\text{Real power mismatch } \Delta P_p = P_p - P_{p\,\text{cal}}$$

$$\text{Reactive power mismatch } \Delta Q_p = Q_p - Q_{p\,\text{cal}}$$

The starting values change from iteration to iteration. Therefore, omitting the superscripts indicating the iteration count, the mismatch equations are

$$\Delta P_p = \sum_{q=2}^{n} \frac{\partial P_p}{\partial \delta_q} \Delta\delta_q + \sum_{q=m+2}^{n} \frac{\partial P_p}{\partial |V_q|} \Delta|V_q|$$

$$\Delta Q_p = \sum_{q=2}^{n} \frac{\partial Q_p}{\partial \delta_q} \Delta\delta_q + \sum_{q=m+2}^{n} \frac{\partial Q_p}{\partial |V_q|} \Delta|V_q|$$

and in matrix form

$$
\begin{pmatrix}
\Delta P_2 \\ \cdots \\ \cdots \\ \cdots \\ \Delta P_n \\ \Delta Q_{m+2} \\ \cdots \\ \cdots \\ \cdots \\ \Delta Q_n
\end{pmatrix}
=
\begin{pmatrix}
\left(\dfrac{\partial P_2}{\partial \delta_2}\right) & \cdots & \left(\dfrac{\partial P_2}{\partial \delta_n}\right) & \left(\dfrac{\partial P_2}{\partial |V_{m+2}|}\right) & \cdots & \left(\dfrac{\partial P_2}{\partial |V_n|}\right) \\
\left(\dfrac{\partial P_2}{\partial \delta_2}\right) & \cdots & \left(\dfrac{\partial P_n}{\partial \delta_n}\right) & \left(\dfrac{\partial P_n}{\partial |V_{m+2}|}\right) & \cdots & \left(\dfrac{\partial P_n}{\partial |V_n|}\right) \\
\left(\dfrac{\partial Q_{m+2}}{\partial \delta_2}\right) & \cdots & \left(\dfrac{\partial Q_{m+2}}{\partial \delta_n}\right) & \left(\dfrac{\partial Q_{m+2}}{\partial |V_{m+2}|}\right) & \cdots & \left(\dfrac{\partial Q_{m+2}}{\partial |V_n|}\right) \\
\left(\dfrac{\partial Q_n}{\partial \delta_2}\right) & \cdots & \left(\dfrac{\partial Q_n}{\partial \delta_n}\right) & \left(\dfrac{\partial Q_n}{\partial |V_{m+2}|}\right) & \cdots & \left(\dfrac{\partial Q_n}{\partial |V_n|}\right)
\end{pmatrix}
=
\begin{pmatrix}
\Delta\delta_2 \\ \cdots \\ \cdots \\ \cdots \\ \Delta\delta_n \\ \Delta|V_{m+2}| \\ \cdots \\ \cdots \\ \cdots \\ \Delta|V_n|
\end{pmatrix}
\tag{4.8}
$$

The matrix containing the differential terms in the above equation is called **Jacobian matrix** and is of $[n - 1 + n - (m + 1)] = (2n - m - 2)$ size and gives the linearised relationship between small changes in voltage angle $\Delta\delta$ and voltage magnitude $\Delta|V|$ with small changes in real and reactive power ΔP and ΔQ. The Jacobian has been partitioned to emphasize the different types of partial derivatives appearing in each sub matrix.

Rewriting Eq. (4.8), we have

$$
\begin{pmatrix} \Delta P \\ \Delta Q \end{pmatrix}
=
\begin{pmatrix}
\left(\dfrac{\partial P}{\partial \delta}\right) & \left(\dfrac{\partial P}{\partial |V|}\right) \\
\left(\dfrac{\partial Q}{\partial \delta}\right) & \left(\dfrac{\partial Q}{\partial |V|}\right)
\end{pmatrix}
\begin{pmatrix} \Delta\delta \\ \Delta|V| \end{pmatrix}
$$

$$
\begin{pmatrix} \Delta P \\ \Delta Q \end{pmatrix}
=
\begin{pmatrix} J_{11} & J_{12} \\ J_{21} & J_{22} \end{pmatrix}
\begin{pmatrix} \Delta\delta \\ \Delta|V| \end{pmatrix}
\tag{4.9}
$$

Equations (4.6) and (4.7) can be rewritten as

$$P_p = V_p^2 Y_{pp} \cos \theta_{pp} + \sum_{\substack{q=1 \\ q \neq p}}^{n} V_p Y_{pq} V_q \cos(\theta_{pq} + \delta_q - \delta_p) \quad \text{where } p = 1, 2, ..., n$$

$$P_p = G_{pp} V_p^2 + \sum_{\substack{q=1 \\ q \neq p}}^{n} V_p Y_{pq} V_q \cos(\theta_{pq} + \delta_q - \delta_p) \quad \text{where } p = 1, 2, ..., n \quad (4.10)$$

since $Y = G + jB$

$$Q_p = -B_{pp} V_p^2 - \sum_{\substack{q=1 \\ q \neq p}}^{n} V_p Y_{pq} V_q \sin(\theta_{pq} + \delta_q - \delta_p) \quad \text{where } p = 1, 2, ..., n \quad (4.11)$$

The diagonal and off-diagonal elements of J_{11} derived from Eq. (4.10) are

$$\frac{\partial P_p}{\partial \delta_p} = \sum_{\substack{q=1 \\ q \neq p}}^{n} V_p Y_{pq} V_q \sin(\theta_{pq} + \delta_q - \delta_p) \quad \text{where } p = 2, ..., n \quad (4.12)$$

$$\frac{\partial P_p}{\partial \delta_q} = -V_p Y_{pq} V_q \sin(\theta_{pq} + \delta_q - \delta_p) \quad \text{where } p = 2, ..., n;\ q = 2, ..., n \neq p \quad (4.13)$$

Similarly, the diagonal and off-diagonal elements of J_{12} are

$$\frac{\partial P_p}{\partial |V_p|} = 2G_{pp} V_p + \sum_{\substack{q=1 \\ q \neq p}}^{n} V_p Y_{pq} V_q \cos(\theta_{pq} + \delta_q - \delta_p) \quad \text{where } p = 2, ..., n \quad (4.14)$$

$$\frac{\partial P_p}{\partial |V_q|} = V_p Y_{pq} \cos(\theta_{pq} + \delta_q - \delta_p) \quad \text{where } p = 2, ..., n;\ q = 2, ..., n \neq p \quad (4.15)$$

The diagonal and off-diagonal elements of J_{21} derived from Eq. (4.11) are

$$\frac{\partial Q_p}{\partial \delta_p} = \sum_{\substack{q=1 \\ q \neq p}}^{n} V_p Y_{pq} V_q \cos(\theta_{pq} + \delta_q - \delta_p) \quad \text{where } p = m + 2, ..., n \quad (4.16)$$

$$\frac{\partial Q_p}{\partial \delta_q} = -V_p Y_{pq} V_q \cos(\theta_{pq} + \delta_q - \delta_p) \quad \text{where } p = m + 2, ..., n;\ q = m + 2, ..., n \neq p \quad (4.17)$$

Similarly, the diagonal and off-diagonal elements of J_{22} are

$$\frac{\partial Q_p}{\partial |V_p|} = -2B_{pp} V_p - \sum_{\substack{q=1 \\ q \neq p}}^{n} Y_{pq} V_q \sin(\theta_{pq} + \delta_q - \delta_p) \quad \text{where } p = m + 2, ..., n \quad (4.18)$$

$$\frac{\partial Q_p}{\partial |V_q|} = -V_p Y_{pq} \sin(\theta_{pq} + \delta_q - \delta_p) \quad \text{where } p = m + 2, ..., n;\ q = m + 2, ..., n \neq p \quad (4.19)$$

4.3.3 Algorithm for Newton–Raphson Method

Step 1 Read system data and convergence criterion (tolerance) ε_P and ε_Q.

Step 2 Formulate $Y_{\text{bus.}}$

Step 3 Select slack bus, if not specified.

Step 4 Assume initial voltages for all the buses (see Figure 4.4);
$$V_p^0 = V_{p\,\text{spec}}\angle 0^\circ \text{ at all generation buses}$$
$$V_p^0 = 1\angle 0^\circ \text{ at all load buses (flat start).}$$

Step 5 Set iteration count $k = 0$.
$$P_p = P_{Gp} - P_{Lp}; \quad Q_p = Q_{Gp} - Q_{Lp};$$
$$Q_{p\,\text{min}} = Q_{Gp\,\text{min}} - Q_{Lp}; \quad Q_{p\,\text{max}} = Q_{Gp\,\text{max}} - Q_{Lp}$$

Step 6 Set bus count $p = 1$.

Step 7 If p is the slack bus, go to Step 13.

Step 8 If p is the load bus, go to Step 10.

Step 9 Compute $Q_p = -\sum_{q=1}^{n} V_p Y_{pq} V_q \sin(\theta_{pq} + \delta_q - \delta_p)$

Check for Q limit violation.

if $Q_{p\,\text{min}} < Q_p < Q_{p\,\text{max}}$, then Go to Step 11

if $Q_p < Q_{p\,\text{min}}$, then $Q_p = Q_{p\,\text{min}}$

if $Q_p > Q_{p\,\text{max}}$, then $Q_p = Q_{p\,\text{max}}$

When Q limit is violated, then the bus is treated as load bus during this iteration;
$V_p = 1\angle 0^\circ$

Step 10 Calculate Q_p, ΔQ_p.
$$Q_p = -\sum_{q=1}^{n} V_p Y_{pq} V_q \sin(\theta_{pq} + \delta_q - \delta_p)$$
$$\Delta Q_p = Q_p - Q_{p\,\text{cal}}$$

Step 11 Calculate P_p, ΔP_p.
$$P_p = \sum_{q=1}^{n} V_p Y_{pq} V_q \cos(\theta_{pq} + \delta_q - \delta_p)$$
$$\Delta P_p = P_p - P_{p\,\text{cal}}$$

Step 12 If p is less than number of buses, increment the bus count and go to Step 7.

Step 13 Check for convergence. If $|\Delta P|_{\text{max}} > \varepsilon_P$ and $|\Delta Q|_{\text{max}} > \varepsilon_Q$, continue. Otherwise go to Step 17.

Step 14 Compute J_{11}, J_{12}, J_{21} and J_{22} and formulate Jacobian matrix.

$$\begin{pmatrix} J_{11} & J_{12} \\ J_{21} & J_{22} \end{pmatrix} = \begin{pmatrix} \dfrac{\partial P}{\partial \delta} & \dfrac{\partial P}{\partial |V|} \\[2mm] \dfrac{\partial Q}{\partial \delta} & \dfrac{\partial Q}{\partial |V|} \end{pmatrix}$$

Step 15   Obtain correction vector

$$\begin{bmatrix} \Delta\delta \\ \Delta V \end{bmatrix} = \begin{bmatrix} J_{11} & J_{12} \\ J_{21} & J_{22} \end{bmatrix}^{-1} \begin{bmatrix} \Delta P \\ \Delta Q \end{bmatrix}$$

Step 16   Update

$$\delta^{k+1} = \delta^k + \Delta\delta^k$$
$$|V^{k+1}| = |V^k| + \Delta|V^k|$$

Increment iteration count and go to Step 6.

Step 17   Calculate slack bus power, reactive power at *PV* buses, line flows and losses.

Flow chart for Newton–Raphson method is shown in Figure 4.4.

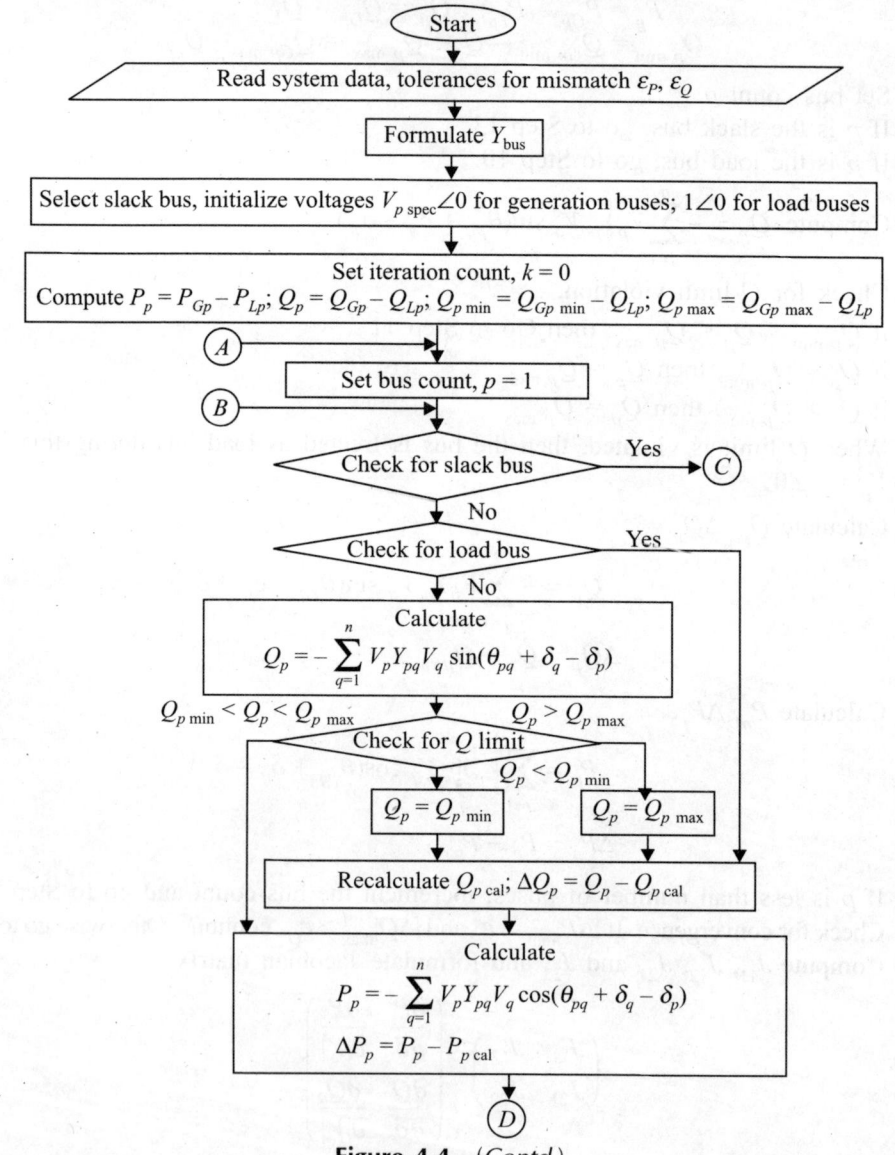

Figure 4.4   (*Contd.*)

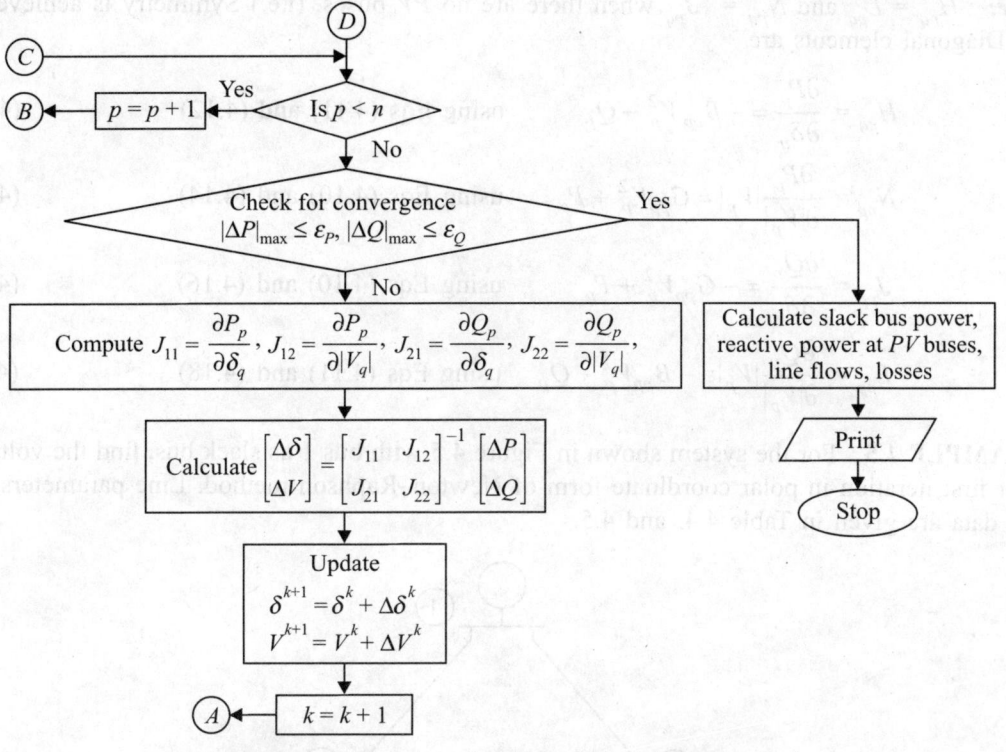

Figure 4.4 Flow chart for Newton–Raphson method.

It may be noted from Eqs. (4.12) to (4.19) that there is no symmetry in Jacobian. Symmetry can be achieved by modifying the independent variable $\Delta|V|$ into $\dfrac{\Delta|V|}{|V|}$ which is a normalized value.

Equation (4.9) becomes

$$\begin{pmatrix} \Delta P \\ \Delta Q \end{pmatrix} = \begin{pmatrix} H & N \\ J & L \end{pmatrix} \begin{pmatrix} \Delta\delta \\ \dfrac{\Delta|V|}{|V|} \end{pmatrix} \tag{4.20}$$

where off-diagonal elements are

$$H_{pq} = \frac{\partial P_p}{\partial \delta_q} = -V_p Y_{pq} V_q \sin(\theta_{pq} + \delta_q - \delta_p) \tag{4.21}$$

$$N_{pq} = \frac{\partial P_p}{\partial |V_q|}|V_q| = V_p Y_{pq} V_q \cos(\theta_{pq} + \delta_q - \delta_p) \tag{4.22}$$

$$J_{pq} = \frac{\partial Q_p}{\partial \delta_q} = -V_p Y_{pq} V_q \cos(\theta_{pq} + \delta_q - \delta_p) \tag{4.23}$$

$$L_{pq} = \frac{\partial Q_p}{\partial |V_q|}|V_q| = -V_p Y_{pq} V_q \sin(\theta_{pq} + \delta_q - \delta_p) \tag{4.24}$$

Note: $H_{pq} = L_{pq}$ and $N_{pq} = -J_{pq}$ when there are no *PV* buses. (i.e.) Symmetry is achieved. Diagonal elements are

$$H_{pq} = \frac{\partial P_p}{\partial \delta_q} = - B_{pp} V_p^2 - Q_p \qquad \text{using Eqs (4.11) and (4.12)} \qquad (4.25)$$

$$N_{pp} = \frac{\partial P_p}{\partial |V_p|} |V_p| = G_{pp} V_p^2 + P_p \qquad \text{using Eqs (4.10) and (4.14)} \qquad (4.26)$$

$$J_{pp} = \frac{\partial Q_p}{\partial \delta_q} = - G_{pp} V_p^2 + P_p \qquad \text{using Eqs (4.10) and (4.16)} \qquad (4.27)$$

$$L_{pp} = \frac{\partial Q_p}{\partial |V_p|} |V_p| = - B_{pp} V_p^2 + Q_p \qquad \text{using Eqs (4.11) and (4.18)} \qquad (4.28)$$

EXAMPLE 4.5 For the system shown in Figure 4.5 with bus 1 as slack bus, find the voltages after first iteration in polar coordinate form of Newton–Raphson method. Line parameters and bus data are given in Table 4.4. and 4.5.

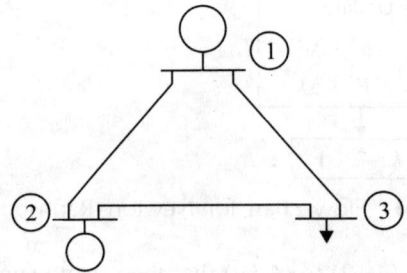

Figure 4.5 System for Example 4.5.

Table 4.4 Line parameters

Bus codes	Line impedance in pu
1–2	$j0.1$
2–3	$j0.2$
3–1	$j0.2$

Table 4.5 Bus data

Bus code	Generation P	Generation Q	Load P	Load Q	$|V|$	Reactive power limit Q_{min}	Q_{max}
1	—	—	—	—	1.0	—	—
2	5	0	—	—	1.1	0	5
3	—	—	3.5	0.5	—	—	—

Solution

Step 1   Line admittances are

$$y_{12} = \frac{1}{j0.1} = - j10$$

$$y_{23} = \frac{1}{j0.2} = - j5$$

$$y_{31} = \frac{1}{j0.2} = - j5$$

Bus admittance matrix is

$$Y_{\text{bus}} = \begin{pmatrix} -j15 & j10 & j5 \\ j10 & -j15 & j5 \\ j5 & j5 & -j10 \end{pmatrix} = \begin{pmatrix} 15\angle-90° & 10\angle90° & j5\angle90° \\ 10\angle90° & 15\angle-90° & j5\angle90° \\ 5\angle90° & 5\angle90° & 10\angle-90° \end{pmatrix}$$

Step 2 Initialise bus voltage and power injections

$$V_1^0 = 1\angle0° \text{ (Slack bus)}$$
$$V_2^0 = 1.1\angle0° \text{ (PV bus)}$$
$$V_3^0 = 1\angle0° \text{ (PQ bus)}$$
$$P_2 = P_{G2} - P_{L2} = 5 - 0 = 5 \text{ pu}$$
$$P_3 = P_{G3} - P_{L3} = 0 - 3.5 = -3.5 \text{ pu}$$
$$Q_3 = Q_{G3} - Q_{L3} = 0 - 0.5 = -0.5 \text{ pu}$$

Step 3 Check for Q limit violation for PV bus

$$Q_p = \sum_{q=1}^{n} V_p Y_{pq} V_q \sin(\theta_{pq} + \delta_q - \delta_p)$$

$$Q_p^0 = -\sum_{q=1}^{n} V_p^0 Y_{pq} V_q^0 \text{ since } (\theta_{pq} + \delta_q^0 - \delta_p^0) = 90° \text{ and } \theta_{pp} = -90°$$

$$Q_2^0 = -[V_2^0 Y_{21} V_1 - (V_2^0)^2 Y_{22} + V_2^0 Y_{23} V_3^0]$$
$$= -[1.1 \times 1 \times 10 - 1.1^2 \times 1.5 + 1.1 \times 1 \times 5]$$
$$= -(11 - 18.15 + 5.5) = 1.65 \text{ pu}$$

Given $0 < Q_2 < 5$

Q_2^0 is within limits. Therefore, Bus 2 acts as PV bus.

$$Q_3^0 = -[V_3^0 Y_1 V_{31} - (V_3^0)^2 Y_{33} + V_3^0 Y_2^0 Y_{23}]$$
$$= -[1 \times 1 \times 5 + 1^2 \times 10 + 1 \times 1.1 \times 5]$$
$$= -(5 - 10 + 5.5) = -0.5 \text{ pu}$$

Step 4 Calculate ΔP and ΔQ

$$P_p = \sum_{q=1}^{n} V_p Y_{pq} V_q \cos(\theta_{pq} + \delta_q - \delta_p)$$

$$P_p^0 = 0 \text{ since } (\theta_{pq} + \delta_q^0 - \delta_p^0) = 90° \text{ and } \theta_{pp} = -90°$$

$$P_2^0 = 0 \text{ and } P_3^0 = 0$$

$$\Delta P_2^0 = P_2 - P_2^0 = 5 - 0 = 5$$

$$\Delta P_3^0 = P_3 - P_3^0 = -3.5 - 0 = -3.5$$

$$\Delta Q_3^0 = Q_3 - Q_3^0 = -0.5 - (-0.5) = 0$$

Step 5 Power flow equation in matrix form is

$$
\begin{pmatrix} \Delta P_2 \\ \Delta P_3 \\ \Delta Q_3 \end{pmatrix} =
\begin{pmatrix}
\dfrac{\partial P_2}{\partial \delta_2} & \dfrac{\partial P_2}{\partial \delta_3} & \dfrac{\partial P_2}{\partial |V_3|} \\[2mm]
\dfrac{\partial P_3}{\partial \delta_2} & \dfrac{\partial P_3}{\partial \delta_3} & \dfrac{\partial P_3}{\partial |V_3|} \\[2mm]
\dfrac{\partial Q_3}{\partial \delta_2} & \dfrac{\partial Q_3}{\partial \delta_3} & \dfrac{\partial Q_3}{\partial |V_3|}
\end{pmatrix}
\begin{pmatrix} \Delta \delta_2 \\ \Delta \delta_3 \\ \Delta |V_3| \end{pmatrix}
$$

$$
\frac{\partial P_p}{\partial \delta_p} = \sum_{\substack{q=1 \\ q \neq p}}^{n} V_p Y_{pq} V_q \, \sin(\theta_{pq} + \delta_q - \delta_p)
$$

$$
\left(\frac{\partial P_p}{\partial \delta_q} \right)^0 = \sum_{\substack{q=1 \\ q \neq p}}^{n} V_p^0 Y_{pq} V_q^0 \text{ since } (\theta_{pq} + \delta_q^0 - \delta_p^0) = 90°
$$

$$
\left(\frac{\partial P_2}{\partial \delta_2} \right)^0 = V_2^0 Y_{12} V_1 + V_2^0 Y_{23} V_3^0 = 11 + 5.5 = 16.5
$$

$$
\left(\frac{\partial P_3}{\partial \delta_3} \right)^0 = V_3^0 Y_{31} V_1 + V_3^0 Y_{32} V_2^0 = 5 + 5.5 = 10.5
$$

$$
\frac{\partial P_p}{\partial \delta_q} = -V_p Y_{pq} V_q \, \sin(\theta_{pq} + \delta_q - \delta_p)
$$

$$
\left(\frac{\partial P_p}{\partial \delta_q} \right)^0 = -V_p Y_{pq} V_q \text{ since } (\theta_{pq} + \delta_q^0 - \delta_p^0) = 90°
$$

$$
\left(\frac{\partial P_2}{\partial \delta_3} \right)^0 = -V_2^0 Y_{23} V_3^0 = -5.5
$$

$$
\left(\frac{\partial P_3}{\partial \delta_3} \right)^0 = -V_3^0 Y_{32} V_2^0 = -5.5
$$

$$
\frac{\partial P_p}{\partial |V_q|} = V_p Y_{pq} \cos(\theta_{pq} + \delta_q - \delta_p)
$$

$$
\left(\frac{\partial P_p}{\partial |V_q|} \right)^0 = 0 \text{ since } (\theta_{pq} + \delta_q^0 - \delta_p^0) = 90°
$$

$$
\left(\frac{\partial P_2}{\partial |V_3|} \right)^0 = 0
$$

$$
\frac{\partial P_p}{\partial |V_p|} = 2 V_p Y_{pp} \cos \theta_{pp} + \sum_{\substack{q=1 \\ q \neq p}}^{n} Y_{pq} V_q \cos(\theta_{pq} + \delta_q - \delta_p)
$$

$$\left(\frac{\partial P_p}{\partial |V_p|}\right)^0 = 0 \text{ since } (\theta_{pq} + \delta_q^0 - \delta_p^0) = 90° \text{ and } \theta_{pp} = -90°$$

$$\left(\frac{\partial P_3}{\partial |V_3|}\right)^0 = 0$$

$$\frac{\partial Q_p}{\partial \delta_p} = \sum_{\substack{q=1 \\ q \neq p}}^{n} V_p Y_{pq} V_q \cos(\theta_{pq} + \delta_q - \delta_p)$$

$$\left(\frac{\partial Q_p}{\partial \delta_p}\right)^0 = 0 \text{ since } (\theta_{pq} + \delta_q^0 - \delta_p^0) = 90°$$

$$\left(\frac{\partial Q_3}{\partial \delta_3}\right)^0 = 0$$

$$\frac{\partial Q_p}{\partial \delta_p} = -V_p Y_{pq} V_q \cos(\theta_{pq} + \delta_q - \delta_p)$$

$$\left(\frac{\partial Q_p}{\partial \delta_q}\right)^0 = 0$$

$$\left(\frac{\partial Q_3}{\partial \delta_2}\right)^0 = 0$$

$$\frac{\partial Q_p}{\partial |V_p|} = -2V_p Y_{pp} \sin \theta_{pp} - \sum_{\substack{q=2 \\ q \neq p}}^{n} Y_{pq} V_q \sin(\theta_{pq} + \delta_q - \delta_p)$$

$$\left(\frac{\partial Q_p}{\partial |V|_p}\right)^0 = 2V_p^0 Y_{pp} - \sum_{\substack{q=2 \\ q \neq p}}^{n} Y_{pq} V_q^0 \text{ since } (\theta_{pq} + \delta_q^0 - \delta_p^0) = 90° \text{ and } \theta_{pp} = -90°$$

$$\left(\frac{\partial Q_3}{\partial |V_3|}\right)^0 = 2V_3^0 Y_{33} - Y_{31} V_1 - Y_{32} V_2^0$$

$$= 2 \times 1 \times 10 - 5 \times 1 - 5 \times 1.1 = 20 - 5 - 5.5$$

$$= 9.5$$

The correction vector is, therefore, calculated from

$$\begin{pmatrix} 5 \\ -3.5 \\ 0 \end{pmatrix} = \begin{pmatrix} 16.5 & -5.5 & 0 \\ -5.5 & 10.5 & 0 \\ 0 & 0 & 9.5 \end{pmatrix} \begin{pmatrix} \Delta\delta_2 \\ \Delta\delta_3 \\ \Delta|V_3| \end{pmatrix}^0 \begin{pmatrix} \Delta\delta_2 \\ \Delta\delta_3 \end{pmatrix}$$

$$\begin{pmatrix} 5 \\ -3.5 \end{pmatrix} = \begin{pmatrix} 16.5 & -5.5 \\ -5.5 & 10.5 \end{pmatrix} \begin{pmatrix} \Delta\delta_2 \\ \Delta\delta_3 \end{pmatrix}$$

$$(0) = (9.5)(\Delta|V_3|)^0$$

$$\Delta |V_3|^0 = \frac{0}{9.5} = 0$$

$$\begin{pmatrix} \Delta\delta_2 \\ \Delta\delta_3 \end{pmatrix}^0 = \begin{pmatrix} 16.5 & -5.5 \\ -5.5 & 10.5 \end{pmatrix}^{-1} \begin{pmatrix} 5 \\ -3.5 \end{pmatrix}$$

$$= \begin{pmatrix} 0.0734 & 0.0385 \\ 0.0385 & 0.1154 \end{pmatrix} \begin{pmatrix} 5 \\ -3.5 \end{pmatrix} = \begin{pmatrix} 0.233 \\ -0.212 \end{pmatrix}$$

$$\delta_2^1 = \delta_2^0 + \Delta\delta_2^0 = 0 + 0.233 = 0.233 \text{ rad} = 13.32°$$

$$\delta_3^1 = \delta_3^0 + \Delta\delta_3^0 = 0 - 0.212 = -0.212 \text{ rad} = -12.12°$$

$$V_3^1 = V_3^0 + \Delta V_3^0 = 1 + 1 = 1 \text{ pu}$$

Note: There is no link between P and $|V|$ and between Q and δ.

4.4   Fast Decoupled Power Flow

It may be observed from the Jacobian Eqs (4.12) to (4.19) that the cross link between ΔP and ΔV and between ΔQ and $\Delta\delta$ is very weak, since

$$\sin(\theta_{pq} + \delta_q - \delta_p) \approx 1$$
$$\cos(\theta_{pq} + \delta_q - \delta_p) \approx 0$$

In other words, the changes which occur in bus active power due to small changes in bus voltage magnitude is very small as compared to their changes due to small changes in bus voltage phase angle. Similarly, the changes which occur in bus reactive power due to small changes in bus voltage phase angle is quite small compared to their changes due to small changes in bus voltage magnitude.

Equation (4.19) therefore becomes decoupled as

$$\begin{pmatrix} \Delta P \\ \Delta Q \end{pmatrix} = \begin{pmatrix} J_{11} & 0 \\ 0 & J_{22} \end{pmatrix} \begin{pmatrix} \Delta\delta \\ \Delta|V| \end{pmatrix}$$

$$\Delta P = J_{11}\Delta\delta = \frac{\partial P}{\partial\delta}\Delta\delta \qquad (4.29)$$

$$\Delta Q = J_{22}\Delta\delta = \frac{\partial Q}{\partial\Delta|V|}\Delta|V|$$

If these equations are used for iteration to find solution, then the method is called *Decoupled method*. For Fast Decoupled method we simplify further.

From Eq. (4.12)

$$\frac{\partial P_p}{\partial\delta_p} = \sum_{\substack{q=1 \\ q \neq p}}^{n} V_p Y_{pq} V_q \sin(\theta_{pq} + \delta_q - \delta_p) \quad \text{where } p = 2, ..., n$$

$$= -Q_p - V_p^2 B_{pp} \text{ [using Eq. (4.11)]}$$

$B_{pp} \gg Q_p$ $\quad \therefore$ We can neglect Q_p

$$V_p^2 \approx V_p$$

$$\frac{\partial P_p}{\partial \delta_p} = -B_{pp}V_p$$

$$\frac{\Delta P_p}{|V_p|} = -B_{pp}\Delta\delta_p \qquad\qquad (4.30)$$

From Eq. (4.13), we get

$$\frac{\partial P_p}{\partial \delta_q} = -V_p Y_{pq} V_q \sin(\theta_{pq} + \delta_q - \delta_p)$$

As $(\delta_q - \delta_p)$ is negligibly small and $V_q \approx 1$,

$$\frac{\partial P_p}{\partial \delta_q} = -V_p Y_{pq} \sin\theta_{pq} = -B_{pp}V_p$$

$$\frac{\Delta P_p}{|V_p|} = -B_{pq}\Delta\delta_p \qquad\qquad (4.31)$$

$$\frac{\Delta P}{|V|} = -B'\Delta\delta$$

$$\Delta\delta = -[B']^{-1}\frac{\Delta P}{|V|} \qquad\qquad (4.32)$$

where B' is the imaginary part of Y_{bus} matrix for all buses except slack bus.
From Eq. (4.18), we get

$$\frac{\partial Q_p}{\partial |V_p|} = -2B_{pp}V_p - \sum_{\substack{q=1 \\ q \neq p}}^{n} Y_{pq}V_q \sin(\theta_{pq} + \delta_q - \delta_p)$$

By multiplying by V_p, we have

$$V_p\frac{\partial Q_p}{\partial |V_p|} = -2B_{pp}V_p^2 - \sum_{\substack{q=1 \\ q \neq p}}^{n} V_p Y_{pq}V_q \sin(\theta_{pq} + \delta_q - \delta_p)$$

$$= -B_{pp}V_p^2 + Q_p \quad [\text{using Eq. (4.11)}]$$

$$= -B_{pp}V_p^2 \quad\quad \text{since } B_{pp} \gg Q_p$$

$$\frac{\partial Q_p}{\partial |V_p|} = -2B_{pp}V_p^2 - \sum_{\substack{q=1 \\ q \neq p}}^{n} V_p Y_{pq}V_q \sin(\theta_{pq} + \delta_q - \delta_p)$$

$$\frac{\partial Q_p}{\partial |V_p|} = -V_p B_{pp}$$

$$\frac{\Delta Q_p}{|V_p|} = -B_{pp}\Delta V_p$$

From Eq. (4.19)

$$\frac{\partial Q_p}{\partial |V_q|} = -V_p Y_{pq} \sin(\theta_{pq} + \delta_q - \delta_p)$$

$$= -V_p Y_{pq} \sin \theta_{pq} \quad \text{since } (\delta_q - \delta_p) \text{ is very small}$$

$$= -B_{pp} V_p$$

or

$$\frac{\Delta Q_p}{|V_p|} = -B_{pp} \Delta V_p$$

$$\frac{\Delta Q}{|V|} = -B'' \Delta V$$

$$\Delta |V| = -[B'']^{-1} \left(\frac{\Delta Q}{|V|} \right) \tag{4.33}$$

where B'' is the imaginary part of Y_{bus} matrix for PQ buses only, that is, excluding slack and voltage controlled buses.

B' and B'' are real and sparse.

Equations (4.32) and (4.33) are solved alternatively always employing the most recent voltage values (phase angle and magnitude). One iteration implies one solution for $(\Delta \delta)$ to update (δ) and then one solution for $\Delta |V|$ to update $|V|$ and it is called $1 - \delta$ and $1 - V$ iteration. Separate convergence tests are applied for the real and reactive power mismatches.

Flow chart for fast decoupled power flow is shown in Figure 4.6.

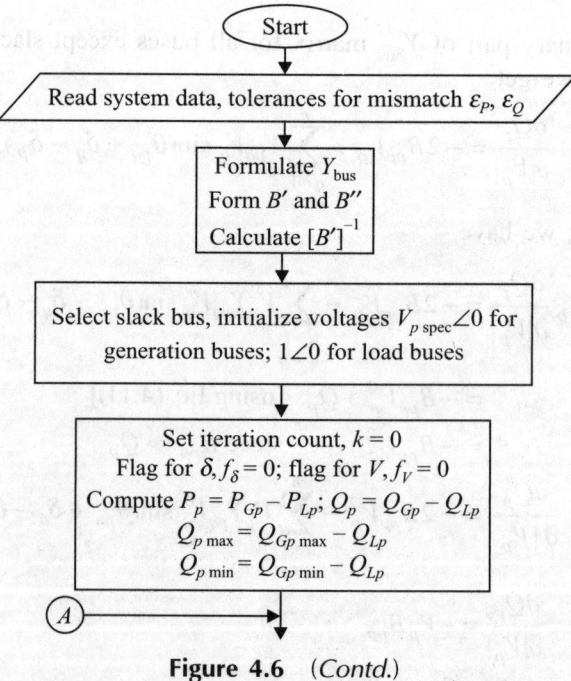

Figure 4.6 (Contd.)

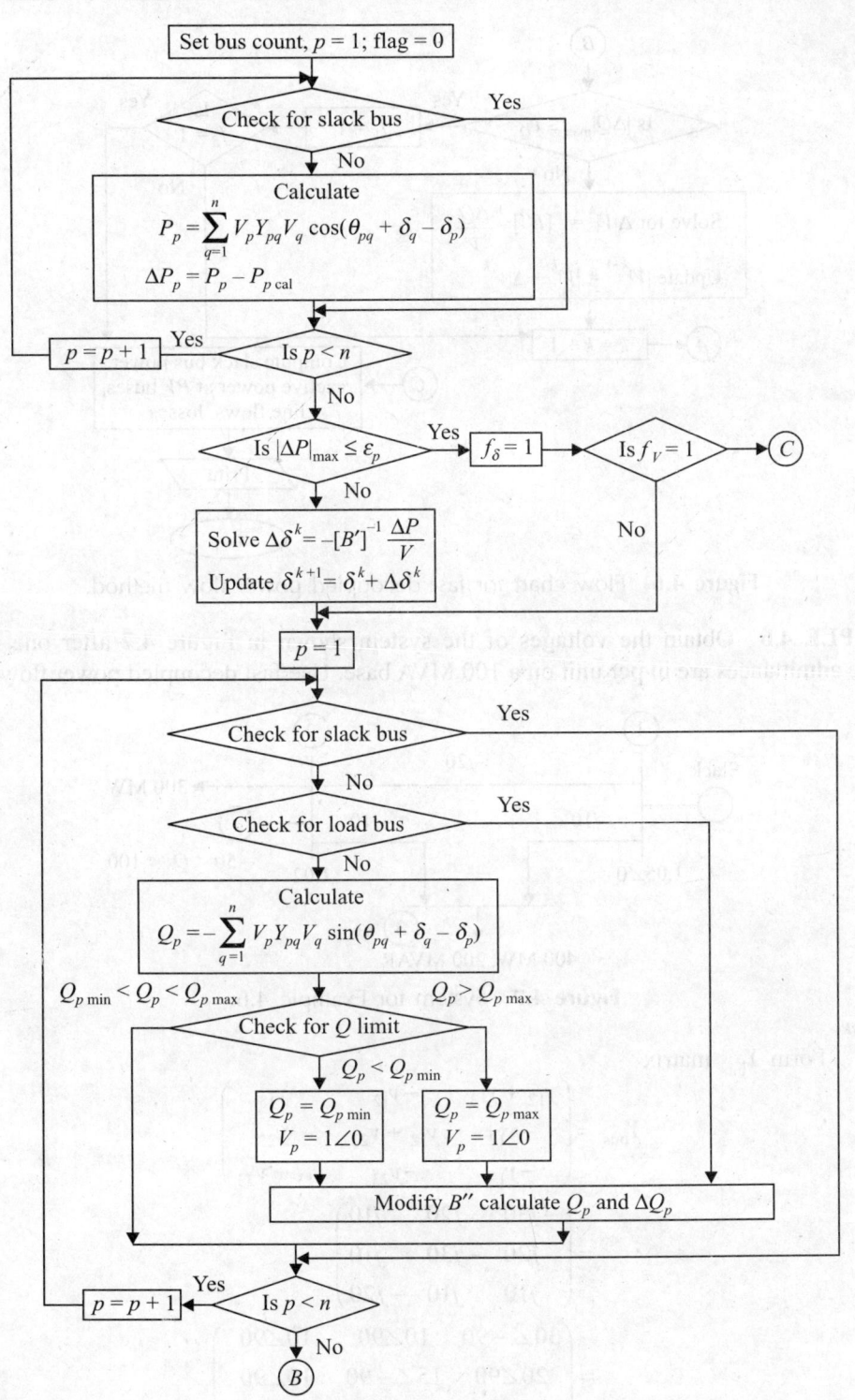

Figure 4.6 (*Contd.*)

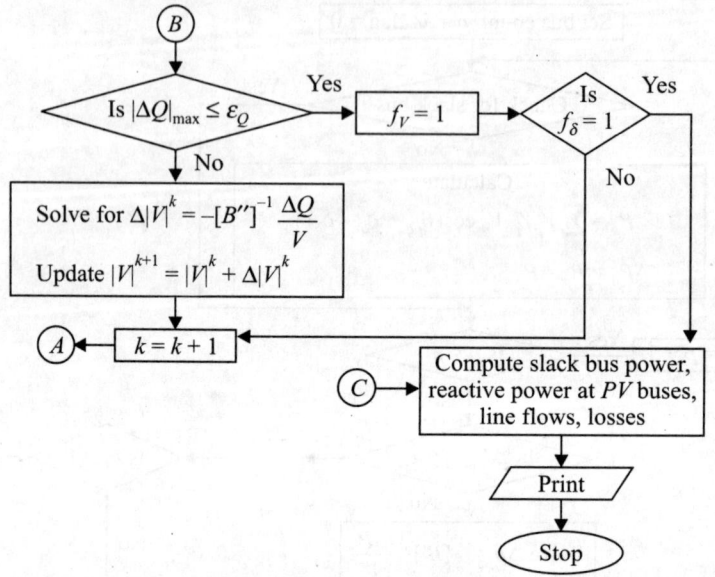

Figure 4.6 Flow chart for fast decoupled power flow method.

EXAMPLE 4.6 Obtain the voltages of the system shown in Figure 4.7 after one iteration. The line admittances are in per unit on a 100 MVA base. Use fast decoupled power flow method.

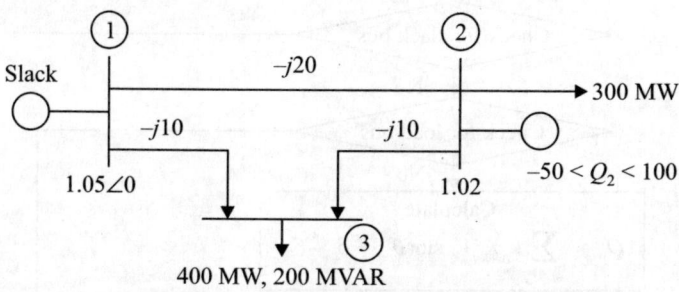

Figure 4.7 System for Example 4.6.

Solution

Step 1   Form Y_{bus} matrix

$$Y_{bus} = \begin{pmatrix} y_{12} + y_{13} & -y_{12} & -y_{13} \\ -y_{12} & y_{12} + y_{23} & -y_{23} \\ -y_{13} & -y_{23} & y_{13} + y_{23} \end{pmatrix}$$

$$= \begin{pmatrix} -j30 & j20 & j10 \\ j20 & -j30 & j10 \\ j10 & j10 & -j20 \end{pmatrix}$$

$$= \begin{pmatrix} 30\angle -90 & 10\angle 90 & 10\angle 90 \\ 20\angle 90 & 15\angle -90 & 10\angle 90 \\ 10\angle 90 & 10\angle 90 & 20\angle -90 \end{pmatrix}$$

$$(B') = \begin{pmatrix} -30 & 10 \\ 10 & -20 \end{pmatrix}$$

$$(B')^{-1} = \frac{1}{500}\begin{pmatrix} -20 & -10 \\ -10 & -30 \end{pmatrix} = -\begin{pmatrix} 0.04 & 0.02 \\ 0.02 & 0.06 \end{pmatrix}$$

Step 2 Initialise bus voltages

$$V_1^0 = 1.05\angle 0° \text{ (Slack bus)}$$

$$V_2^0 = 1.02\angle 0° \text{ (}PV \text{ bus)}$$

$$V_3^0 = 1\angle 0° \text{ (}PQ \text{ bus)}$$

Step 3

$$P_2 = P_{G2} - P_{L2} = \left(\frac{300}{100}\right) - 0 = 3 \text{ pu}$$

$$P_3 = P_{G3} - P_{L3} = 0 - \left(\frac{400}{100}\right) = -4 \text{ pu}$$

$$Q_3 = Q_{G3} - Q_{L3} = 0 - \left(\frac{200}{100}\right) = -2 \text{ pu}$$

$$Q_{2\,min} = Q_{G2\,min} - Q_{L2} = -\left(\frac{50}{100}\right) - 0 = -0.5 \text{ pu}$$

$$Q_{2\,max} = Q_{G2\,max} - Q_{L2} = \left(\frac{100}{100}\right) - 0 = 1 \text{ pu}$$

Step 4

$$P_p = \sum_{q=1}^{n} V_p Y_{pq} V_q \cos(\theta_{pq} + \delta_q - \delta_p)$$

$$P_p^0 = 0 \quad \text{since } \cos(\theta_{pq} + \delta_q^0 - \delta_p^0) = 0$$

$$P_2^0 = 0 \text{ and } P_3^0 = 0$$

$$\Delta P_2^0 = P_2 - P_2^0 = 3$$

$$\Delta P_3^0 = P_3 - P_3^0 = -4$$

Step 5

$$\begin{pmatrix} \Delta\delta_2 \\ \Delta\delta_3 \end{pmatrix}^0 = -(B')^{-1}\begin{pmatrix} \dfrac{\Delta P_2}{|V_2|} \\ \dfrac{\Delta P_3}{|V_3|} \end{pmatrix}^0$$

$$= \begin{pmatrix} 0.04 & 0.02 \\ 0.02 & 0.06 \end{pmatrix}\begin{pmatrix} \dfrac{3}{1.02} \\ \dfrac{-4}{1} \end{pmatrix} = \begin{pmatrix} 0.034 \\ -0.183 \end{pmatrix}$$

$$\delta_2^1 = \delta_2^0 + \Delta\delta_2^0 = 0 + 0.034 = 0.034 \text{ rad} = 1.95°$$

$$\delta_3^1 = \delta_3^0 + \Delta\delta_3^0 = 0 - 0.183 = -0.183 \text{ rad} = -10.49°$$

Step 6

$$Q_p = -V_p^2 Y_{pp} \sin\theta_{pp} - \sum_{\substack{q=1 \\ q\neq p}}^{n} V_p Y_{pq} V_q \sin(\theta_{pq} + \delta_q - \delta_p)$$

$$Q_p^0 = (V_p^0)^2 Y_{pp} - \sum_{\substack{q=1 \\ q\neq p}}^{n} V_p^0 Y_{pq} V_q^0 \sin(\theta_{pq} + \delta_q^1 - \delta_p^1)$$

$$Q_2^0 = 1.02^2 \times 30 - [1.02 \times 1.05 \times 20 \sin(90 + 0 - 1.95) + 1.02 \times 1 \times 10 \sin(90 - 10.49 - 1.95)]$$

$$= 31.21 - (21.42 \times 0.9994 + 10.2 \times 0.9765)$$

$$= -0.16$$

Q_2^0 is within limits.

∴   Bus 2 acts as *PV* bus

∴ $[B''] = [-20]$

$$Q_3^0 = 1^2 \times 20 - [1 \times 1.05 \times 10 \sin\{(90 - 0 - (-10.49)\} + 1 \times 1.02 \times 10 \sin\{90 + 1.95 - (-10.49)\}]$$

$$= 20 - (10.5 \times 0.9833 + 10.2 \times 0.9765)$$

$$= -0.28$$

$$\Delta Q_3^0 = Q_3 - Q_3^0$$

$$= -2 - (-0.28)$$

$$= -1.72$$

Step 7

$$\Delta V_3^1 = -[B'']^{-1}\left(\frac{\Delta Q_3^0}{|V_3^0|}\right)$$

$$= \frac{1}{20}\left(\frac{-1.72}{1}\right) = -0.086$$

$$V_3^1 = V_3^0 + \Delta V_3^1 = 1 - 0.086 = 0.914 \text{ pu}$$

EXAMPLE 4.7   Same as Example 4.6 except for Q limits which are $-5 < Q_2 < 100$ MVAR.

Solution

Steps 1–5 are same as in previous example.

Step 6 $Q_2^0 = -0.16$ pu

But Q_2 limits are $-0.05 < Q_2 < 1$ pu

Since $Q_2^0 < Q_{2\,min}$, consider $Q_2^0 = Q_{2\,min} = -0.05$ pu

Bus 2 will act as *PQ* bus.   ∴ $V_2^0 = 1$ and $[B''] = [B']$.

$$Q_2^0 = 1.02^2 \times 30 - [1.02 \times 1.05 \times 20 \sin(90 + 0 - 1.95) + 1.02 \times 1$$
$$\times 10 \sin(90 - 10.49 - 1.95)]$$
$$= 30 - (21.42 \times 0.9994 + 10.2 \times 0.9765) = -1.37$$
$$Q_3^0 = 1^2 \times 20 - [1 \times 1.05 \times 10 \sin\{(90 - 0 - (-10.49)\} + 1 \times 1.02$$
$$\times 10 \sin\{90 + 1.95 - (-10.49)\}]$$
$$= 20 - (10.5 \times 0.9833 + 10.2 \times 0.9765) = -0.28$$
$$\Delta Q_2^0 = Q_2 - Q_2^0 = -0.05 - (-1.37) = 1.32$$
$$\Delta Q_3^0 = Q_3 - Q_3^0 = -2 - (-0.28) = -1.72$$

Step 7

$$\begin{pmatrix} |\Delta V_2| \\ |\Delta V_3| \end{pmatrix}^1 = -(B'')^{-1} \begin{pmatrix} \dfrac{\Delta Q_2}{|V_2|} \\ \dfrac{\Delta Q_3}{|V_3|} \end{pmatrix}^0 = -\begin{pmatrix} 0.04 & 0.02 \\ 0.02 & 0.06 \end{pmatrix} \begin{pmatrix} \dfrac{1.32}{1} \\ \dfrac{-1.72}{1} \end{pmatrix}$$

$$= -\begin{pmatrix} 0.0528 - 0.0344 \\ 0.0264 - 0.1032 \end{pmatrix} = \begin{pmatrix} -0.0184 \\ 0.0768 \end{pmatrix}$$

$$V_2^1 = V_2^0 + \Delta V_2^1 = 1 - 0.0184 = 0.9816 \text{ pu}$$
$$V_3^1 = V_3^0 + \Delta V_3^1 = 1 + 0.0768 = 1.0768 \text{ pu}$$

The various methods of solving power flow solution are compared in the following Table 4.6.

Table 4.6 Comparison of power flow methods

Attribute	Gauss–Seidal (GS)	Newton–Raphson (NR)	FDPF
Reliability	Less accurate and less reliable	More accurate and more reliable	More reliable
Speed of convergence	• Less no. of arithmetic operation to complete one iteration and therefore less time per iteration. • More number of iterations to obtain solution • Linear convergence characteristic. So slow rate of convergence. • Convergence affected by ill-conditioned system-series capacitors	• More time per iteration as Jacobian is to be computed in each iteration. • Less number of iterations. • Quadratic convergence characteristic. So speed of convergence is fast. • Convergence not affected by ill-conditioned system.	• Less time per iteration • More number of iterations. • Geometric convergence characteristic. So speed of convergence is fast.
Dependency on system size	Number of iterations increases with increase in size of system.	Number of iterations does not depend on system size.	
Simplicity of programming	Programming is simple.	Programming is complex.	Programming is simple.

(Contd.)

Table 4.6 Comparison of power flow methods (*Contd.*)

Attribute	Gauss–Seidal (GS)	Newton–Raphson (NR)	FDPF
Memory requirement	Less	More	Less than NR method
Selection of slack bus	Convergence is affected by selection of slack bus.	Convergence is not affected by selection of slack bus.	
Application	Suitable for small system	Suitable for large system	Suitable for contingency evaluation for security assessment and for solving optimization problems
Coordinate	Rectangular coordinate	Both rectangular and polar coordinates	Polar coordinate

Short Answer Questions

1. What are the iterative methods mainly used to solve power flow problems?
   The iterative methods mainly used to solve power flow problems are
   - Gauss–Seidal method
   - Newton–Raphson method
   - Fast Decoupled Power Flow method

2. When is the generator bus treated as load bus?
   The generation bus is treated as load bus when the reactive power of the generation violates the specified limits. The reactive power is pegged to the limit it has violated and the voltage of the bus is assumed as $1 + j0$.

3. Why is the acceleration factor used in Gauss–Seidal method of power flow solution algorithm?
   The number of iterations in Gauss–Seidal method of power flow solution algorithm can be reduced if the voltage at the load buses is multiplied by acceleration factor, bringing the voltages closer to the final values. Convergence is thus speeded up by using the acceleration factor.

4. What are the facts based on which the approximation of Newton–Raphson method to decoupled power flow method is made?
   The facts are:
   (i) Real power at a bus does not change appreciably for a small change in the bus magnitude.
   (ii) Reactive power at a bus does not change appreciably for a small change in the bus phase angle.

5. Why is the time required per iteration more in the Newton–Raphson method?
   Due to the need for evaluating the elements of the Jacobian in each iteration, the time required per iteration is more in the Newton–Raphson method.

6. Why is Y_{bus} used in power flow solution instead of Z_{bus}?
   The advantages of using bus admittance matrix for power flow studies are
   (i) Data preparation is simple.
   (ii) Bus admittance matrix can be easily formed and modified for network changes such as addition or deletion of lines, transformers, etc.
   (iii) Unlike bus impedance matrix which is a full matrix, bus admittance matrix is highly sparse and therefore reduces computer memory and execution time.

7. What is the major disadvantage of decoupled power flow method?
   It requires more number of iterations for accurate solution because of the approximations.

8. What are the major advantages of FDPF over Newton–Raphson method?
   FDPF is superior to Newton–Raphson method from the points of view of storage and speed.

9. Compare the Gauss–Seidal and Newton–Raphson methods of power flow solution.
   Refer Table 4.6.

10. Compare the FDPF and Newton–Raphson methods of power flow solution.
    Refer Table 4.6.

11. What is a swing bus or slack bus?
    The swing bus or slack bus is a generation bus whose voltage magnitude and phase angle are specified. The losses corresponding to the difference between generation and load are supplied by this bus.

Exercises

4.1 Figure E4.1 shows the one line diagram of a simple three bus power system with generation at bus 1. The scheduled loads on buses 2 and 3 are marked on the diagram. Line reactances are marked in per unit on a 100 MVA base.

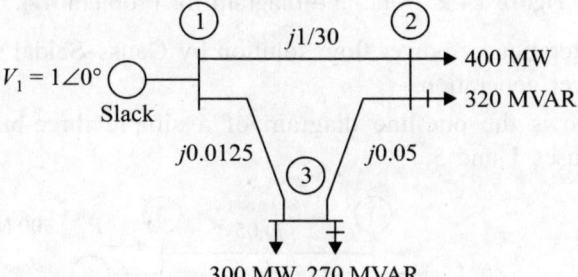

Figure E4.1   One-line diagram for Problem 4.1.

(a) Using Gauss–Seidal method determine V_2 and V_3 after one iteration.
(b) After several iterations the bus voltages converge to

$$V_2 = 0.9 - j0.1 \text{ pu}$$
$$V_3 = 0.95 - j0.05 \text{ pu}$$

Determine the line flows and line losses and the slack bus real and reactive power. Construct a power flow diagram and show the direction of the line flows.

[**Ans.** $V_2^1 = 0.9360 - j0.0800$

$\qquad V_3^1 = 0.9602 - j0.0460$

$\qquad S_{12} = 300$ MW $+ j300$ MVAR; $\qquad S_{21} = -300$ MW $- j240$ MVAR

$\qquad S_{L12} = j60$ MVAR

$\qquad S_{13} = 400$ MW $+ j400$ MVAR; $\qquad S_{31} = -400$ MW $- j360$ MVAR

$\qquad S_{L13} = j40$ MVAR

$\qquad S_{23} = -100$ MW $- j80$ MVAR; $\qquad S_{32} = 100$ MW $+ j90$ MVAR

$\qquad S_{L23} = j10$ MVAR

$\qquad S_1 = 700$ MW $+ j700$ MVAR]

4.2 Do Problem 4.1 using Newton–Raphson method.

4.3 Do Problem 4.1 using FDPF method.

4.4 Figure E4.2 shows a three-bus power system.

Bus 1: Slack bus, $V = 1.05\angle0°$ pu

Bus 2: $\quad PV$ bus, $V = 1.0$ pu, $\quad P_G = 3$pu

Bus 3: $\quad PQ$ bus, $P_L = 4$ pu, $\quad Q_L = 2$ pu.

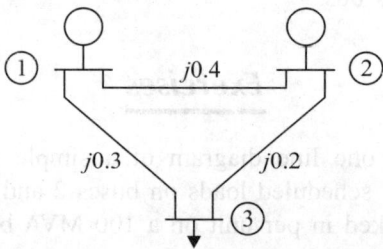

Figure E4.2   One-line diagram for Problem 4.4.

Carry out one iteration of power flow solution by Gauss–Seidal method. Neglect limits on reactive power generation.

4.5 Figure E4.3 shows the one-line diagram of a simple three-bus power system with generation of buses 1 and 3.

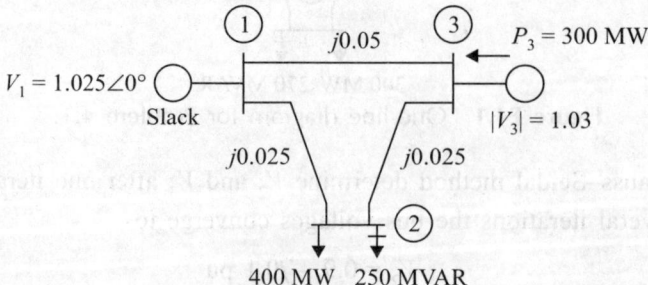

Figure E4.3   One-line diagram for Problem 4.5.

(a) Using Gauss–Seidal method determine V_2 and V_3 after one iteration.

(b) After several iterations the bus voltages converge to

$$V_2 = 1.0012\angle{-2.1°}$$
$$V_3 = 1.03\angle{1.37°}$$

Determine the line flows and line losses and the slack bus real and reactive power. Construct a power flow diagram and show the direction of the line flows.

[**Ans.**

$$V_2^1 = 1.0025 - j0.05; \qquad\qquad Q_3^1 = 1.236$$

$$V_3^1 = 1.0299 + j0.0152$$

$$S_{12} = 150.4 \text{ MW} + j100.2 \text{ MVAR}; \quad S_{21} = -150.4 \text{ MW} - j92 \text{ MVAR}$$

$$S_{L12} = j8.2 \text{ MVAR}$$

$$S_{13} = -50.4 \text{ MW} - j9.6 \text{ MVAR}; \quad S_{31} = 50.4 \text{ MW} + j10.9 \text{ MVAR}$$

$$S_{L13} = j1.3 \text{MVAR}$$

$$S_{23} = -249.6 \text{ MW} - j107.6 \text{ MVAR}; \quad S_{32} = 249.6 \text{ MW} + j126 \text{ MVAR}$$

$$S_{L23} = j18.4 \text{ MVAR}$$

$$S_1 = 100 \text{ MW} + j90.5 \text{ MVAR]}$$

4.6 Do Problem 4.5 using Newton–Raphson method.

4.7 Do Problem 4.5 using FDPF method.

4.8 Figure E4.4 shows the one-line diagram of a simple three-phase power system with generation at buses 1 and 2. Determine the bus admittance matrix. Using Gauss–Seidal method determine the voltages V_2 and V_3 after one iteration.

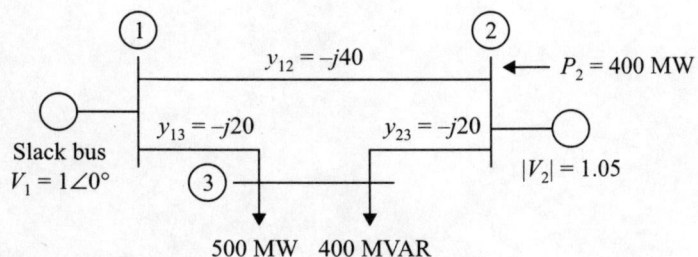

Figure E4.4   One-line diagram for Problem 4.8.

$$\textbf{Ans.} \quad \begin{bmatrix} -j60 & j40 & j20 \\ j40 & -j60 & j20 \\ j20 & j20 & -j40 \end{bmatrix}$$
$$V_2 = 1.05\angle{1.58°}, V_3 = 0.92\angle{-6.18°}$$

4.9 Do Problem 4.8 using FDPF method.   [**Ans.** $V_2 = 1.05\angle{1.5°}; \; V_3 = 0.925\angle{-6.41°}$]

4.10 The following is the data for a power flow solution

Bus code	Admittance
1–2	$2 - j8$
1–3	$1 - j3$
2–3	$0.6 - j2$
2–4	$1 - j4$
3–4	$2 - j8$

The schedule of active and reactive powers is

Bus code	P	Q	V	Remarks
1	—	—	$1.05 + j0$	Slack
2	0.5	0.2	$1 + j0$	PQ
3	0.4	0.3	$1 + j0$	PQ
4	0.3	0.1	$1 + j0$	PQ

Determine the voltage at the end of first iteration using Gauss–Seidal method. Take acceleration factor = 1.4.

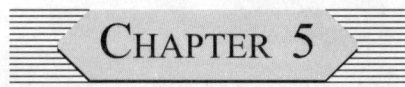

CHAPTER 5

Balanced Fault Analysis

The normal mode of operation of a power system is balanced on all the three phases. A number of undesired but unavoidable disturbances can temporarily disrupt this condition. Such a disturbance is said to be a fault. A fault in a power system may be caused by the following reasons:

- Insulation failure
- Swinging of lines
- Snapping of conductor
- Shorting of a line to ground or other line by birds
- Falling of a tree or branch over the line
- Falling of towers due to vehicle collision or hurricane.

The fault is the result of one or more conductors coming in contact with ground or one another. The potential difference causes the fault current, whose magnitude and severity depends on the path it takes.

5.1 Need for Fault Analysis

Fault analysis is performed to
- determine bus voltages and line currents during various types of faults
- select suitable rating for machines, transformers, lines/cables, switch gear (including fuses and circuit breakers) and relay setting
- design the grounding system
- provide appropriate cable joints.

5.2 Classification of Faults

Faults are generally classified into two categories—symmetrical or balanced faults, and unsymmetrical or unbalanced faults.

5.2.1 Symmetrical Faults

In symmetrical faults, all the three phases are short circuited, and voltages and currents remain balanced even after the fault. It is, therefore, amenable for per-phase analysis.

A three-phase fault occurs rarely but it is the most severe type of fault that disturbs the system to the maximum extent.

5.2.2 Unsymmetrical Faults

In unsymmetrical faults, one or two phases are involved. Voltages and currents become unbalanced and each phase is to be treated separately. Depending on the phases involved, the faults can be classified as

- Line-to-ground (LG) fault
- Line-to-line (LL) fault
- Line-to-line-to-ground (LLG) or double line-to-ground fault.

The relative frequencies of occurrence of faults are about

3-phase	3%
LLG	7%
LL	10%
LG	80%

The relative frequency also indicates their effect on the system in the order of decreasing severity on the system.

5.3 Modelling

Accurate fault calculations involve much labour and time, and in many a situation it is also not required. Hence, the following assumptions are made:

1. Synchronous machine is modelled as constant voltage behind a reactance—subtransient reactance is considered for determining fault current immediately after the fault and transient reactance is used for determining fault current after about three cycles.
2. Shunt elements are neglected in the representation of transformer and transmission line. Load impedances are also neglected.
3. Series resistance of generator and transformer are neglected. Sometimes series resistance of line is also neglected.
4. Transformer tap is at nominal value.

In normal operation, the bus voltages are maintained at the nominal value. All pre-fault voltages are, therefore, taken as $1\angle 0°$. If it is not so, pre-fault voltages are computed from power flow studies.

Further the pre-fault current, even if it flows, is very small compared to the fault current, and can be neglected. If the pre-fault current is to be considered when calculating fault current, then superposition principle can be applied.

5.4 Thevenin's Theorem and Application

The current flowing immediately after the fault consists of an ac component which eventually reaches steady state and a fast decaying dc component which decays to zero as shown in Figure 5.1.

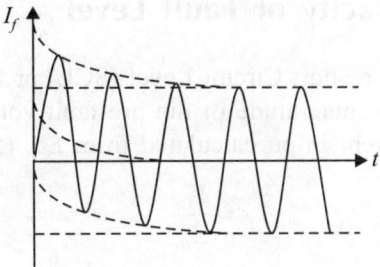

Figure 5.1 Fault current.

For analysis only the ac component is considered as shown in Figure 5.2.

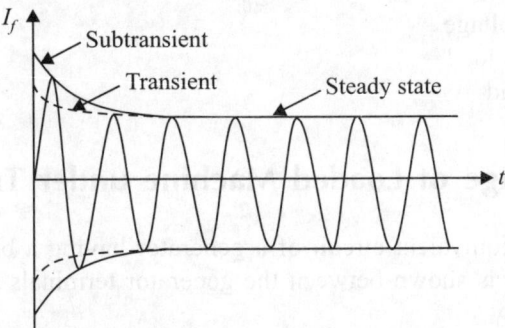

Figure 5.2 The ac component of fault current.

A linear, bilateral and passive network can be replaced at any pair of terminals a and b, by an equivalent network consisting of a voltage source V_{th} in series with an impedance Z_{th}, where V_{th} and Z_{th} are computed by opening the terminals a and b, and Z_{th} is the network impedance as seen from the terminals a and b with all the sources replaced by their internal impedances.

V_{th} is called the **Thevenin voltage** and Z_{th} is the **Thevenin impedance**. If the fault occurs across a–b with a fault impedance of z_f, then the equivalent circuit is shown in Figure 5.3 and the fault current is given by

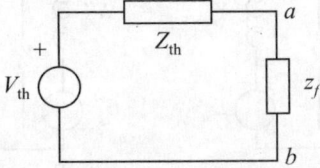

Figure 5.3 Thevenin equivalent circuit with fault impedance.

$$I_f = \frac{V_{th}}{Z_{th} + z_f} \qquad (5.1)$$

If the fault impedance is zero, then the fault is called **bolted** or **solid fault**.

5.5   Short Circuit Capacity or Fault Level

Short Circuit Capacity (SCC) or Short Circuit Level (SCL) or fault level at a bus of a network is defined as the product of the magnitude of the pre-fault voltage and the fault current.

For a solid fault, fault current in pu calculated from Eq. (5.1) is

$$I_f = \frac{V_{th}}{Z_{th}}$$

$$SCC = V_{th} I_f = \frac{V_{th}^2}{Z_{th}}$$

$\Rightarrow$
$$SCC = \frac{1}{Z_{th}} \text{ in pu} \qquad (5.2)$$

since V_{th} = pre-fault voltage
$$= 1 \text{ pu on no load}$$
$$\approx 1 \text{ pu on load}$$

5.6   Internal Voltage of Loaded Machine under Transient Conditions

Figure 5.4(a) shows the equivalent circuit of a generator having a balanced three-phase motor load. External impedance is shown between the generator terminals and the point F where the fault occurs.

The current flowing before the fault occurs at F is I_L, the voltage V_f and the generator terminal voltage V_t. The equivalent circuit of the synchronous generator is its no load voltage E_g in series with its direct axis synchronous reactance X_{dg}. If a three-phase fault occurs at F in the system, the conditions for calculating subtransient or transient current are not satisfied since the reactance of the generator must be either X_{dg}'' or X_{dg}'.

The circuit shown in Figure 5.4(b) gives the desired result. Here a voltage E_g'' in series with X_{dg}'' supplies the steady state current I_L when switch S is open and supplies the current to the

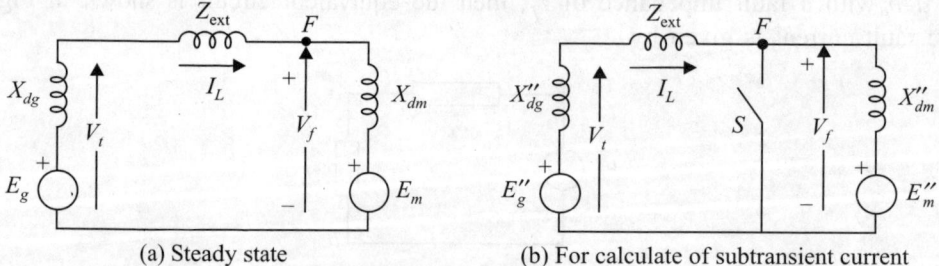

(a) Steady state   (b) For calculate of subtransient current
Figure 5.4   Equivalent circuit for a generator supplying a balanced three-phase motor load.

short circuit through X''_{dg} and Z_{ext} when switch S is closed. If E''_g is determined as follows when S is open,

$$E''_g = V_t + jI_L X''_{dg} \tag{5.3}$$

then when S is closed the subtransient fault current will be $I''_f \cdot E''_g$ is called the subtransient internal voltage. Similarly, when calculating transient current I'_f which must be supplied through the transient reactance X'_{dg}, the driving voltage is the transient internal voltage E'_g given by

$$E'_g = V_t + jI_L X'_{dg} \tag{5.4}$$

Voltages E''_g and E'_g are determined by I_L and are both equal to the no load voltage E_g only when I_L is zero, at which time E_g is equal to V_t.

It is to be noted that E''_g in series with X''_{dg} represents the generator before the fault occurs and immediately after the fault, only if the pre-fault current is I_L. On the other hand, E_g in series with the direct axis synchronous reactance X_{dg} is the equivalent circuit of the machine under steady state conditions for any load. For a different value of I_L in the circuit of Figure 5.4(a), E_g would remain the same with different V_t, but a new value of E''_g would be required.

Synchronous motors have reactances of the same type as generators. When a motor is short circuited, it no longer receives electrical energy from the power system, but its field remains energized and the inertia of its rotor and connected load keeps it rotating. The internal voltage of the synchronous motor contributes current to the fault, for it is then acting like a generator. By comparing with the corresponding formulas for a generator, the subtransient and transient internal voltages for a synchronous motor are given by

$$E''_m = V_t - jI_L X''_{dm} \tag{5.5}$$
$$E'_m = V_t - jI_L X'_{dm} \tag{5.6}$$

Systems that contain generators and motors under load may be solved by Thevenin's theorem or by the use of circuit analysis.

5.7 Procedure for Calculating Fault or Short-Circuit Current

Step 1   Draw the single line diagram.
Step 2   Select a common base and convert all impedances to per unit values on this base.
Step 3   Draw the impedance diagram.
Step 4   Calculate by circuit analysis the fault current and currents from machines.

or

Step 1   Calculate Thevenin impedance, which is the impedance seen from the fault point by circuit analysis. This may involve series parallel combination, star-delta or delta-star transformation.
Step 2   Calculate Thevenin voltage, which is the pre-fault voltage at fault point from power flow solution.

Step 3   Determine fault current and fault level.

Step 4   Current from each machine is computed using superposition principle.

$$I_{f\,mach} = I_{f\,contribution\ by\ mach} + I_L \tag{5.7}$$

where I_L is pre-fault current flowing through the machine.

5.8   Calculation of Fault Current using Circuit Analysis

The fault current may be computed using the principles of circuit theory. The following examples describe the methodology.

EXAMPLE 5.1   Two generators are connected in parallel to the low voltage side of a three phase Δ–Y transformer as shown in Figure 5.5. Generator G_1 is rated 50 MVA, 13.8 kV. Generator G_2 is rated 25 MVA, 13.8 kV. Each generator has a subtransient reactance of 25%. The transformer is rated 75 MVA, 13.8 Δ/69Y kV, with a reactance of 10%. Before the fault occurs, the voltage on the high tension side of the transformer is 66 kV. The transformer is unloaded, and there is no circulating current between the generators. Find the subtransient current in each generator when a three phase short circuit occurs on the high tension side of the transformer.

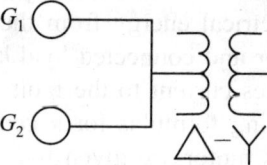

Figure 5.5   One-line diagram for Example 5.1.

Solution   Since fault current from generator is required, select as base 13.8 kV and 75 MVA in the generator side (low tension circuit). Then the base voltage on the high tension side is $13.8 \times (69/13.8) = 69$ kV.

G_1:
$$X''_d = 0.25 \times \left(\frac{75}{50}\right) = 0.375 \text{ pu}$$

$$E_{g1} = 66 \times \left(\frac{13.8}{69}\right) \Big/ 13.8 = 0.957 \text{ pu}$$

G_2:
$$X''_d = 0.25 \times \left(\frac{75}{25}\right) = 0.75 \text{ pu}$$

$$E_{g2} = 66 \times \left(\frac{13.8}{69}\right) \Big/ 13.8 = 0.957 \text{ pu}$$

Transformer:
$$X_T = 0.1 \text{ pu}$$

Figure 5.6(a) shows the reactance diagram before the faults.

The internal voltages of the two machines may be considered to be in parallel since they must be identical in magnitude and phase if no circulating current flows between them.

They can be replaced by an equivalent single machine as shown in Figure 5.6(b).

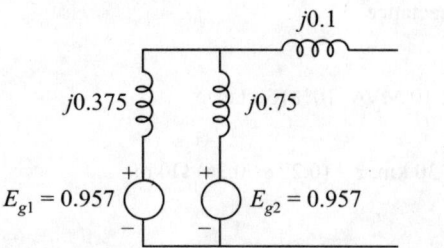

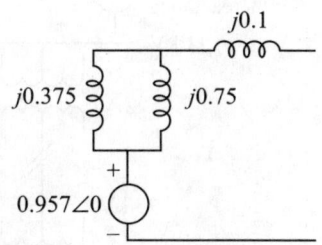

Figure 5.6(a) Reactance diagram for Example 5.1.

Figure 5.6(b) Simplified reactance diagram for Example 5.1.

The subtransient short circuit current is

$$I_f'' = \frac{0.957\angle 0}{(j0.375 \parallel j0.75) + j0.1} = \frac{0.957\angle 0}{j0.25 + j0.1}$$

$$= -j2.735 \text{ pu}$$

$$I_{g1}'' = I_f'' \frac{X_{d2}''}{X_{d1}'' + X_{d2}''} = -j2.735 \times \frac{j0.75}{j0.375 + j0.75}$$

$$= -j1.823 \text{ pu}$$

Similarly, $I_{g2}'' = I_f'' - I_{g1}'' = -j2.735 - (-j1.823) = -j0.912 \text{ pu}$

The current contribution from the generators may also be calculated by finding voltage across their subtransient reactances and dividing by their respective substransient reactances as given below.

Voltages on the low-tension side of the transformer is

$$V_T = I_f'' X_T = -j2.735 \times j0.1 = 0.2735 \text{ pu}$$

$$I_{f1}'' = \frac{E_{g2} - V_T}{X_{d2}''} = \frac{0.957 - 0.2735}{j0.375} = -j1.823 \text{ pu}$$

$$I_{f2}'' = \frac{E_{g2} - V_T}{X_{d2}''} = \frac{0.957 - 0.2735}{j0.75} = -j0.912 \text{ pu}$$

which are the same as calculated previously.

$$I_f'' = 2.735 \times 3.138 = 8.58 \text{ kA}$$
$$I_{g1}'' = 1.823 \times 3.138 = 5.72 \text{ kA}$$
$$I_{g2}'' = 0.912 \times 3.138 = 2.86 \text{ kA}$$

EXAMPLE 5.2   For the radial network shown in Figure 5.7, a three-phase fault occurs at F. Determine the fault current and the line voltage at 11 kV bus under fault conditions.

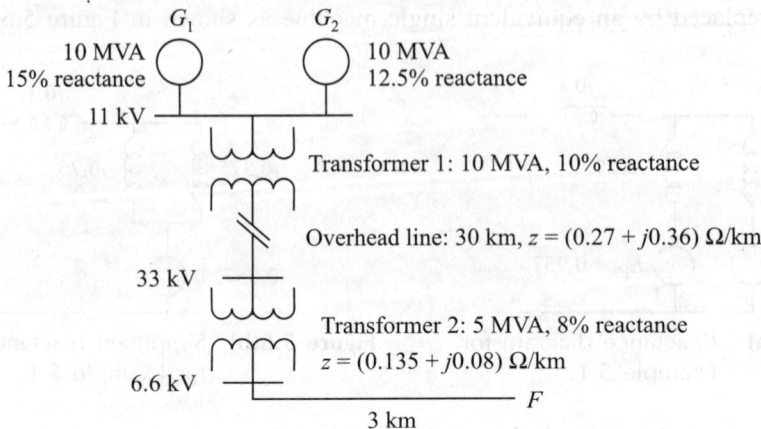

G_1 10 MVA   15% reactance   11 kV

G_2 10 MVA   12.5% reactance

Transformer 1: 10 MVA, 10% reactance

Overhead line: 30 km, $z = (0.27 + j0.36)$ Ω/km

33 kV

Transformer 2: 5 MVA, 8% reactance $z = (0.135 + j0.08)$ Ω/km

6.6 kV

3 km F

Figure 5.7   Radial network for Example 5.2.

Solution

$$\text{Select system base, } S_b = 10 \text{ MVA}$$
$$\text{Voltage bases are} = 11 \text{ kV in generator}$$
$$= 33 \text{ kV for overhead line}$$
$$= 6.6 \text{ kV for cable}$$
$$\text{Reactance of } G_1 = 0.15 \text{ pu}$$
$$\text{Reactance of } G_2 = 0.125 \text{ pu}$$
$$\text{Reactance of } T_1 = 0.1 \text{ pu}$$
$$\text{Reactance of } T_2 = 0.08 \times \left(\frac{10}{5}\right) = 0.16 \text{ pu}$$
$$\text{Impedance of overhead line} = 30 \times (0.27 + j0.36) \times \left(\frac{10}{33^2}\right)$$
$$= 0.0744 + j0.099 \text{ pu}$$
$$\text{Impedance of cable} = 3 \times (0.135 + j0.08) \times \left(\frac{10}{6.6^2}\right)$$
$$= 0.093 + j0.055 \text{ pu}$$

Impedance diagram is shown in Figure 5.8.

Since the system is on no load prior to occurrence of fault and assuming no circulating current between the generators, the voltages of the two generators are identical (in magnitude and phase) and are assumed equal to 1 pu and reference phases. The generators circuit can thus be replaced by a single voltage source in series with the parallel combination of generator reactances as given.

$$\text{Total impedance} = (j0.15 \parallel j0.125) + j0.1 + 0.0744 + j0.099 + j0.16 + 0.093 + j0.055$$
$$= 0.1674 + j0.482 = 0.51\angle 70.8° \text{ pu}$$
$$I_f = \frac{1\angle 0°}{0.51\angle 70.8°} = 1.96\angle -70.8° \text{ pu}$$
$$I_b = \frac{S_b}{\sqrt{3}\,V_b} = \frac{10}{\sqrt{3}\times 6.6} = 0.875 \text{ kA}$$
$$I_f = 1.96 \times 0.875 = 1.715 \text{ kA}$$

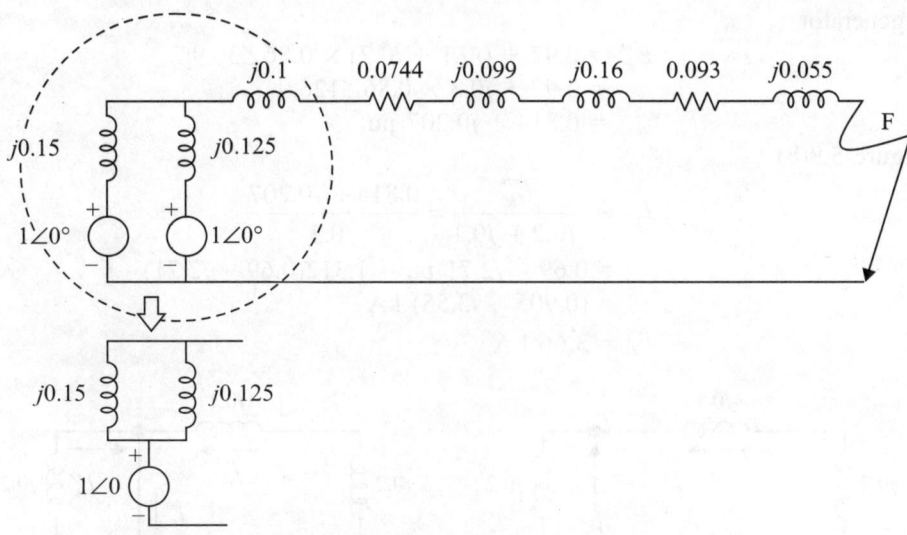

Figure 5.8 Impedance diagram for Example 5.2.

Total impedance between 11 kV bus and fault point F

$$= j0.1 + 0.0744 + j0.099 + j0.16 + 0.093 + j0.055$$
$$= 0.1674 + j0.414 = 0.443\angle 67.8° \text{ pu}$$

Voltage at 11 kV bus $= 1.96\angle -70.8° \times 0.443\angle 67.8°$
$$= 0.88\angle -3° \text{ pu}$$
$$= 0.88 \times 11 = 9.68 \text{ kV}$$

In Examples 5.1 and 5.2 the system was on no load before the occurrence of the fault. Now, we shall consider a case with loaded system on fault.

EXAMPLE 5.3 A synchronous generator and motor are rated 30 MVA, 13.2 kV and both have subtransient reactances of 20%. The line connecting them has a reactance of 10% on the base of the machine ratings. The motor is drawing 20 MW at 0.8 power factor leading and a terminal voltage of 12.8 kV when a symmetrical three-phase fault occurs at the motor terminals. Find the subtransient current in the generator, motor and fault by using the internal voltages of the machines.

Solution Choose as base S_b = 30 MVA and V_b = 13.2 kV.

$$\text{Base current, } I_b = \frac{30}{\sqrt{3} \times 13.2} = 1.312 \text{ kA}$$

Figure 5.9(a) shows the equivalent circuit of the system described.
Choosing the voltage at the fault point F as reference phasor,

$$V_f = \frac{12.8}{13.2} = 0.97\angle 0°$$

$$I_L = \frac{S_L\angle + \varphi}{V_L} = \frac{20/0.8/30}{0.97}\angle \cos^{-1} 0.8°$$
$$= 0.86\angle 36.9° \text{ pu}$$

For the generator

$$E''_g = 0.97 + (j0.1 + j0.2) \times 0.86\angle 36.9°$$
$$= 0.97 + j0.3 \times 0.86\angle 126.9°$$
$$= 0.814 + j0.207 \text{ pu}$$

From Figure 5.9(b),

$$I''_g = \frac{E''_g}{j0.2 + j0.1} = \frac{0.814 + j0.207}{j0.3}$$
$$= 0.69 - j2.71 \text{ pu} = 1.312(0.69 - j2.71)$$
$$= (0.905 - j3.55) \text{ kA}$$
$$I''_g = 3.66 \text{ kA}$$

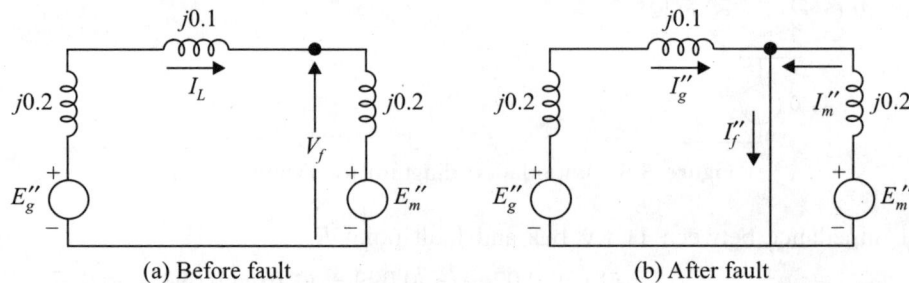

(a) Before fault   (b) After fault

Figure 5.9 Equivalent circuits for Example 5.3.

Similarly for the motor,

$$E''_m = 0.97 - (j0.2) \times 0.86\angle 36.9°$$
$$= 1.074 - j0.138 \text{ pu}$$
$$I''_m = \frac{1.074 - j0.138}{j0.2} = -0.69 - j5.37$$
$$= 1.312(-0.69 - j5.37) = (0.905 - j7.05) \text{ kA}$$
$$I''_m = 7.2 \text{ kA}$$

In the fault

$$I''_f = I''_g + I''_m = 0.69 - j2.71 - 0.69 - j5.37$$
$$= -j8.08 \text{ pu}$$
$$= 1.312(-j8.08) = -j10.6 \text{ kA}$$
$$I''_f = 10.6 \text{ kA}$$

5.9 Calculation of Fault Current using Thevenin's Theorem and Superposition Theorem

The subtransient current in the fault can be found directly by Thevenin's theorem, which is applicable to linear, bilateral circuits. When constant values are used for the reactances of synchronous machines, linearity is assumed. When the theorem is applied to the circuit of Figure 5.10(a), the equivalent circuit is a single generator and single impedance terminating at the point of application of the fault. The new generator has an internal voltage equal to V_f, the voltage at the fault point before the fault occurs. The impedance is that measured at the fault point looking back into the circuit with all the generated voltages replaced by their internal impedances. Subtransient reactances should be used if the initial current is desired. Figure 5.10(b) shows the Thevenin equivalent of Figure 5.10(b).

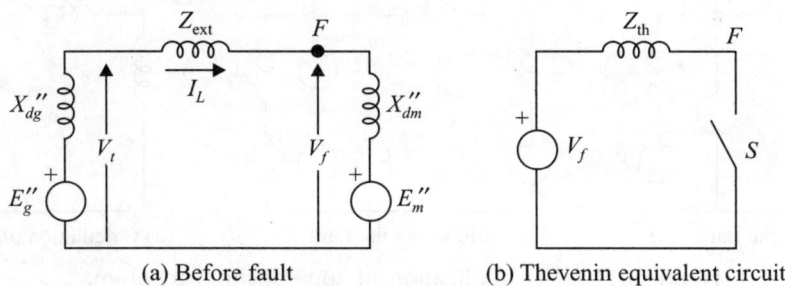

(a) Before fault   (b) Thevenin equivalent circuit

Figure 5.10   Thevenin's method.

The impedance $Z_{th} = (Z_{ext} + jX_d'') \parallel Z_L$

Upon the occurrence of a three phase short circuit at F, simulated by closing S, the subtransient fault current is

$$I_f'' = \frac{V_f}{Z_{th}} = V_f \frac{(Z_L + Z_{ext} + jX_d'')}{Z_L(Z_{ext} + jX_d'')} \tag{5.8}$$

Figure 5.11(a) shows a generator having a voltage V_f connected at the fault and equal to the voltage at the fault before the fault occurs. This generator has no effect on the current flowing before the fault occurs and the circuit corresponds to that of Figure 5.10(a). Adding in series with V_f another generator having an emf of equal magnitude but 180° out of phase with V_f gives the circuit of 5.11(b), which corresponds to that of Figure 5.10(b) with switch S closed.

The principle of superposition is applied by first shorting E_g'', E_m'' and V_f as shown in Figure 5.11(c), which gives the fault current as well as currents through the machines found by distributing the fault current between the two machines inversely as the impedances of their circuits.

$$Z_{th} = \frac{jX_{dm}''(Z_{ext} + jX_{dg}'')}{jX_{dm}'' + Z_{ext} + jX_{dg}''}$$

$$I_f'' = \frac{V_f}{Z_{th} + z_f}$$

where $Z_f = 0$ for solid or bolted fault.

$$\text{Fault current from generator} = I_f'' \times \frac{jX_{dm}''}{jX_{dm}'' + Z_{ext} + jX_{dg}''}$$

$$\text{Fault current from motor} = I''_f \times \frac{Z_{ext} + jX''_{dg}}{jX''_{dm} + Z_{ext} + jX''_{dg}}$$

Then shorting the remaining generator—V_f with E''_g, E''_m and V_f intact in the circuit as shown in Figure 5.11(a) gives the current flowing before the fault. Adding the two values of current in each branch using superposition theorem gives the current in the branch after the fault.

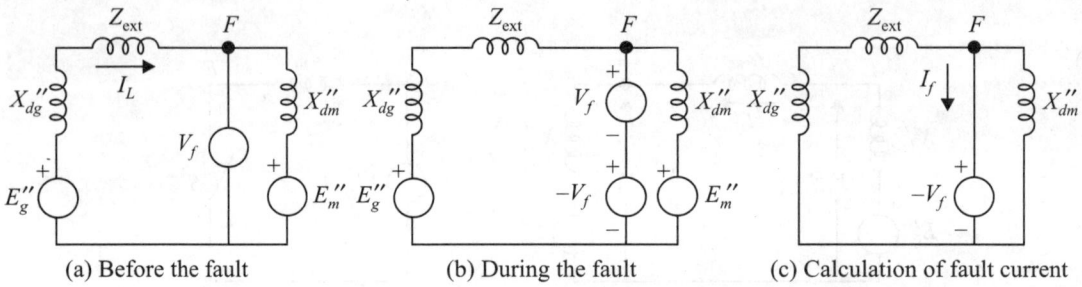

(a) Before the fault   (b) During the fault   (c) Calculation of fault current

Figure 5.11   The application of superposition theorem.

The current in the fault is the same whether or not load current is considered, but the contributions from the lines differ. Usually, load current is omitted in determining the current in each line upon the occurrence of a fault as it is very less compared to fault current contribution. In the Thevenin's method, neglecting load current means that the pre-fault current in each line is zero. If the subtransient internal voltages of all machines are assumed equal to the voltage V_f at the fault point before the fault occurs, then no current flows anywhere in the network prior to the fault.

EXAMPLE 5.4   Solve Example 5.3 by the use of Thevenin's and superposition theorems.

Solution   From Figure 5.12, we have

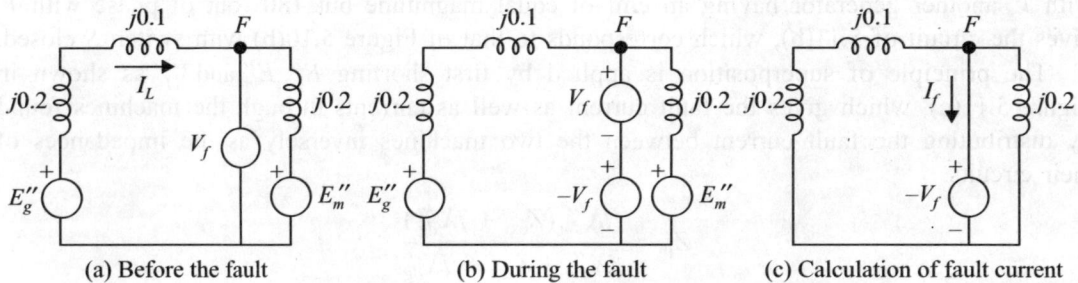

(a) Before the fault   (b) During the fault   (c) Calculation of fault current

Figure 5.12   Circuits for Example 5.4.

$$\text{Fault current from generator} = -j8.08 \times \frac{j0.2}{j0.5} = -j3.23 \text{ pu}$$

$$\text{Fault current from motor} = -j8.08 \times \frac{j0.3}{j0.5} = -j4.85 \text{ pu}$$

To these currents the pre-fault current I_L is added to obtain the total subtransient currents in the machines.

$$I_L = \frac{S_L \angle + \varphi}{V_L} = \frac{20/0.8/30}{0.97} \angle \cos^{-1} 0.8° = 0.86 \angle 36.9° \text{ pu}$$

$$= 0.69 + j0.52 \text{ pu}$$

$$I_g'' = -j3.23 + 0.69 + j0.52 = 0.69 - j2.71 \text{ pu}$$

$$I_m'' = -j4.85 - 0.69 - j0.52 = -0.69 - j5.37 \text{ pu}$$

Note that I_L is in the same direction as I_g'' but opposite to I_m''.

$$I_b = \frac{30}{\sqrt{3} \times 13.2} = 1.312 \text{ kA}$$

$$I_g'' = 1.312(0.69 - j2.71) = (0.905 - j3.55) \text{ kA}$$

$$I_g'' = 3.66 \text{ kA}$$

$$I_m'' = 1.312(-0.69 - j5.37) = (0.905 - j7.05) \text{ kA}$$

$$I_m'' = 7.2 \text{ kA}$$

$$I_f'' = 1.312(-j8.08) = -j10.6 \text{ kA}$$

$$I_f'' = 10.6 \text{ kA}$$

5.10 Selection of Circuit Breaker

The subtransient current discussed so far does not include the dc component. Inclusion of the dc component results in an rms value of current immediately after the fault which is higher than the symmetrical subtransient current and is about 1.6 times the symmetrical value. The breaker must withstand this momentary current which causes disruptive mechanical forces.

The breaker must be capable of interrupting the current after about three cycles from the time of fault by which time the machine reactance would be transient reactance and not the subtransient reactance. The interrupting current is, therefore, lower than the momentary current with the dc component also decaying to lower value. The interrupting current is about 1.2 times the symmetrical transient current.

EXAMPLE 5.5 Figure 5.13(a) shows a 25 MVA, 13.8 kV generator with $X_d'' = 15\%$ connected through a transformer to a bus which supplies four identical motors.

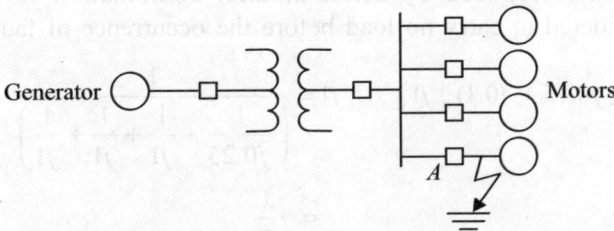

Figure 5.13(a) One-line diagram for Example 5.5.

The subtransient reactance X_d'' of each motor is 20% on a base of 5 MVA, 6.9 kV. The three phase rating of the transformer is 25 MVA, 13.8/6.9 kV, with a leakage reactance of 10%. The bus voltage at the motors is 6.9 kV when a three-phase fault occurs at the point F. For the fault specified, determine the subtransient current in the fault, the subtransient current in breaker A and the symmetrical short circuit interrupting current in the fault and in breaker A ($X_d' = 1.5\,X_d''$).

Solution

$$S_b = 25 \text{ MVA}, \quad V_b = 6.9 \text{ kV on the motor circuit.}$$

$$= 13.8 \times \frac{6.9}{13.8} = 6.9 \text{ kV in the motor circuit}$$

$$I_b = \frac{25}{\sqrt{3} \times 6.9} = 2.09 \text{ kA}$$

$$X_{dg}'' = 0.15, \quad X_t = 0.1 \quad \text{and} \quad X_{dm}'' = 0.2 \times \frac{25}{5} = 1 \text{ pu}$$

Figure 5.13(b) shows the impedance diagram.

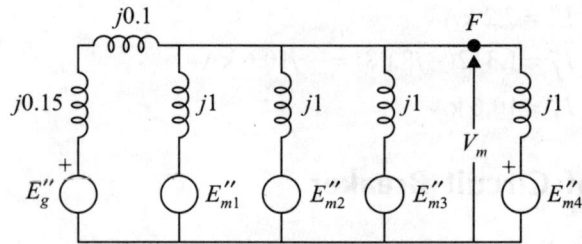

Figure 5.13(b) Impedance diagram for Example 5.5.

For a fault at F,

$$V_f = V_m = \frac{6.9}{6.9} = 1 \text{ pu}$$

The solution can be obtained either by (i) circuit theory and network reduction (ii) Thevenin's theorem.

(i) Circuit theory

Through breaker A, the flow is contributed by the generator and 3 of the 4 motors. Accordingly the reactance diagram is reduced by series-parallel combination to Figure 5.14(a)–(b), as the system is considered to carry no load before the occurrence of fault.

$$(j0.15 + j0.1) \| j1 \| j1 \| j1 = \cfrac{1}{\left(\cfrac{1}{j0.25} + \cfrac{1}{j1} + \cfrac{1}{j1} + \cfrac{1}{j1} \right)}$$

$$= j\frac{1}{7}$$

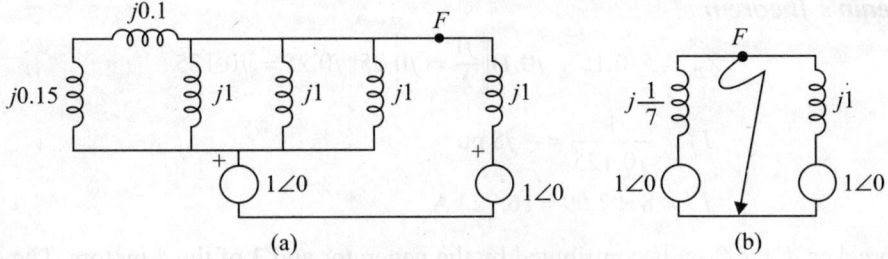

Figure 5.14   Reduced reactance diagram.

Fault current through breaker A, motor 4 and fault are given by

$$I''_{fA} = \frac{1}{\left(j\dfrac{1}{7}\right)} = -j7 \text{ pu}$$

$$I''_{fm4} = \frac{1}{j1} = -j1 \text{ pu}$$

$$I''_f = I''_{fA} + I''_{fm4} = -j8 \text{ pu}$$

$$I''_{fA} = 7 \times 2.09 = 14.63 \text{ kA}$$

$$I''_f = 8 \times 2.09 = 16.72 \text{ kA}$$

To compute the interrupting current by breaker A, replace the subtransient reactance by transient reactance.

$$X'_{dg} = 0.15 \times 1.5 = 0.225$$

$$X'_{dm} = 1 \times 1.5 = 1.5$$

Impedance up to fault location excluding motor 4 is

$$(j0.225 + j0.1) \| j1.5 \| j1.5 \| j1.5 = \frac{1}{\left(\dfrac{1}{j0.325} + \dfrac{1}{j1.5} + \dfrac{1}{j1.5} + \dfrac{1}{j1.5}\right)}$$

$$= j\frac{1}{5.077}$$

$$I'_{fA} = \frac{1}{\left(j\dfrac{1}{5.077}\right)} = -j5.077 \text{ pu}$$

$$I'_{fm4} = \frac{1}{j1.5} = -j0.667 \text{ pu}$$

$$I'_f = I'_{fA} + I'_{fm4} = -j5.744 \text{ pu}$$

$$I'_{fA} = 5.077 \times 2.09 = 10.61 \text{ kA}$$

$$I'_f = 5.744 \times 2.09 = 12 \text{ kA}$$

(ii) Thevenin's theorem

$$Z_{th} = (j0.15 + j0.1)\left\|\frac{j1}{4}\right. = j0.25\|j0.25 = j0.125$$

$$I_f'' = \frac{1}{j0.125} = -j8 \text{ pu}$$

$$I_f'' = 8 \times 2.09 = 16.72 \text{ kA}$$

Through breaker A, the flow is contributed by the generator and 3 of the 4 motors. The generator contributes a current of

$$-j8 \times \frac{0.25}{0.5} = -j4 \text{ pu}$$

Each motor contributes 25% of the remaining fault current or $-j1$ pu.

Through breaker A,

$$I_{fA}'' = -j4 + 3(-j1) = -j7 \text{ pu}$$

$$I_{fA}'' = 7 \times 2.09 = 14.63 \text{ kA}$$

To compute the interrupting current by breaker A, replace the subtransient reactance by transient reactance.

$$X_{dg}' = 0.15 \times 1.5 = 0.225$$

$$X_{dm}' = 1 \times 1.5 = 1.5$$

$$Z_{th} = (j0.225 + j1)\left\|\frac{j1.5}{4}\right.$$

$$= \frac{j0.325 \times j0.375}{j0.700} = j0.174 \text{ pu}$$

$$I_f' = \frac{V_f}{Z_{th}} = \frac{1}{j0.174} = -j5.744 \text{ pu}$$

The generator contributes a current of

$$-j5.744 \times \frac{j0.375}{j0.7} = -j3.077 \text{ pu}$$

Each motor contributes a current of

$$\frac{1}{4}(-j5.744) \times \frac{j0.375}{j0.7} = -j0.667 \text{ pu}$$

The symmetrical short circuit current to be interrupted is

$$(3.077 + 3 \times 0.667) \times 2.09 = 10.61 \text{ kA}$$

The usual procedure is to rate all the breakers connected to a bus on the basis of the current into a fault on the bus. The short circuit current interrupting rating of the breakers connected to the 6.9 kV bus must be at least

$$5.744 \text{ pu} = 5.744 \times 2.09 = 12 \text{ kA}$$

5.11   Series Reactors

With system expansion, sometimes the fault level may increase beyond the short circuit capacity of the breakers. Under the circumstances, series reactor may be used to increase the impedance thereby reducing the fault level within the breaker capacity limits.

EXAMPLE 5.6   A station operating at 33 kV is divided into two sections A and B. Section A consists of three generators 15 MVA each having a reactance of 15% and section B is fed from a grid through a 75 MVA transformer of 8% reactance, as shown in Figure 5.15. The circuit breakers have each a rupturing capacity of 750 MVA. Determine the reactance of the reactor to prevent the breakers being overloaded if a symmetrical short circuit occurs on an outgoing feeder connected to Section A.

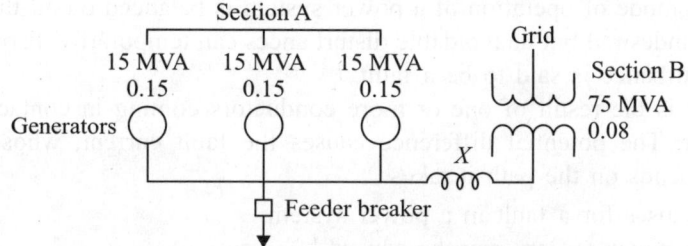

Figure 5.15   System for Example 5.6.

Solution

$$S_b = 75 \text{ MVA}, V_b = 33 \text{ kV} \qquad \text{SCL} = \frac{750}{75} = 10 \text{ pu}$$

$$X_G = 0.15 \times \frac{75}{15} = 0.75 \text{ pu}$$

The impedance diagram is shown in Figure 5.16.

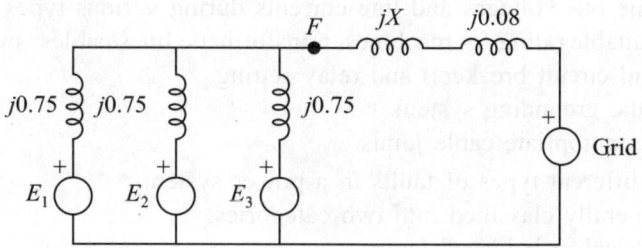

Figure 5.16   Impedance diagram for Example 5.6.

$$\text{SCL} = \frac{1}{X_{\text{th}}} = \frac{1}{X + 0.08} + \frac{1}{0.75} + \frac{1}{0.75} + \frac{1}{0.75} = 10$$

$$\frac{1}{X + 0.08} + 4 = 10$$

$$X + 0.08 = \frac{1}{6} = 0.167$$

$$X = 0.087 \text{ pu}$$

$$\text{Base impedance } Z_b = \frac{33^2}{75}$$

$$X = 0.087 \times \frac{33^2}{75} = 1.26 \, \Omega$$

Short Answer Questions

1. What is meant by fault in a power system?

   The normal mode of operation of a power system is balanced on all the three phases. A number of undesired but unavoidable disturbances can temporarily disrupt this condition. Such a disturbance is said to be a fault.

   The fault is the result of one or more conductors coming in contact with ground or one another. The potential difference causes the fault current, whose magnitude and severity depends on the path it takes.

2. Name the causes for a fault in a power system.

   A fault in a power system may be caused by
   • Insulation failure
   • Swinging of lines
   • Snapping of conductor
   • Shorting a line to ground or other line by birds
   • Falling of a tree or branch over the line
   • Falling of towers due to vehicle collision or hurricane

3. What is the need for fault analysis in a power system?

   Fault analysis is performed to
   (i) determine bus voltages and line currents during various types of faults.
   (ii) select suitable rating for machines, transformers, lines/cables, switch gear (including fuses and circuit breakers) and relay setting
   (iii) design the grounding system
   (iv) provide appropriate cable joints.

4. Classify the different types of faults in a power system.

   Faults are generally classified into two categories:
   (i) Symmetrical or balanced faults
   (ii) Unsymmetrical or unbalanced faults

5. Define symmetrical faults.

   In symmetrical faults, all the three phases are short circuited, and voltages and currents remain balanced even after the fault. That is the magnitude of fault current is equal in all the phases. It is, therefore, amenable for per-phase analysis.

6. Define unsymmetrical faults.

   In unsymmetrical faults, one or two phases are involved. Voltages and currents become unbalanced and each phase is to be treated separately, as the magnitude of fault current is not equal in all the phases. Depending on the number of phases involved, the faults can be classified as

   (i) Line-to-ground (LG) fault

  (ii) Line-to-line (LL) fault

 (iii) Line-to-line-to-ground (LLG) or double line-to-ground fault

7. Write the relative frequencies of occurrence of various types of faults.

The relative frequencies of occurrence of faults are about

3-phase fault	3%
LLG fault	7%
LL fault	10%
LG fault	80%

8. Rank the various types of faults in the order of severity.

The various types of faults in the order of decreasing severity are

3-phase fault

LLG

LL fault

LG fault

9. What is the reason for transients during short circuits?

The short circuits or faults cause sudden change in currents. Most of the components of the power system have inductive property which opposes any sudden change in currents and so the short circuits are associated with transients.

10. How the circuit breakers are selected?

The major specifications for circuit breakers are:

- Normal operating voltage
- Normal operating (rated) current
- Momentary short-circuit current (subtransient)
- Interrupting short-circuit current (transient)
- Speed of opening

11. Define synchronous reactance.

The synchronous reactance of a machine is the ratio of induced emf to the steady state rms current. It is the reactance of a synchronous machine under steady state condition equal to the sum of leakage reactance and reactance representating armature reaction.

$$X_d = X_1 + X_a$$

12. Define subtransient reactance.

The subtransient reactance of a machine is the ratio of induced emf on no load to the subtransient symmetrical rms current. It is the reactance of a synchronous machine under subtransient condition.

$$X_d'' = X_1 + X_a \| X_f \| X_{ad}$$

13. Define transient reactance.

The transient reactance of a machine is the ratio of induced emf on no load to the transient symmetrical rms current. It is the reactance of a synchronous machine under transient condition.

$$X_d' = X_1 + X_a \| X_f$$

14. What is the significance of subtransient reactance in fault analysis?

The subtransient reactance is used to estimate the initial value of fault current immediately on the occurrence of the fault, which causes disruptive mechanical forces. The momentary short-circuit current rating of the circuit breaker should be more than this value.

15. What is the significance of transient reactance in fault analysis?

The transient reactance is used to estimate the fault current during transient period when the circuit breaker interrupts this current. The breaking short-circuit current rating of the circuit breaker should be more than this value.

Exercises

5.1 The system shown in Figure E5.1 is initially on no load with generators operating at their rated voltage with their emfs in phase. The rating of the generators and transformers, and their respective per cent reactances are given in the diagram. The line reactance is 160 Ω. A three-phase fault occurs at the receiving end of the transmission line. Determine the short-circuit current and the short-circuit MVA.

[**Ans.** $288.675\angle-90°$ A, 200 MVA]

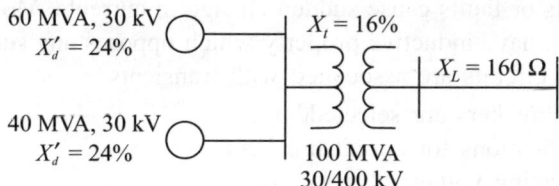

Figure E5.1   One-line diagram for Problem 5.1.

5.2 The system shown in Figure E5.2 shows an existing plant consisting of a generator 100 MVA, 30 kV with 20% subtransient reactance and a generator of 50 MVA, 30 kV with 10% subtransient reactance, connected in parallel to a 30 kV bus bar. The 30 kV bus bar feeds a transmission line via the circuit breaker C which is rated at 1250 MVA. A grid supply is connected to the station bus bar through a 500 MVA, 400/30 kV transformer with 20% reactance. Determine the reactance of a current limiting reactor X in ohm to be connected between the grid system and the existing bus bar such that the short-circuit MVA of the breaker C does not exceed.

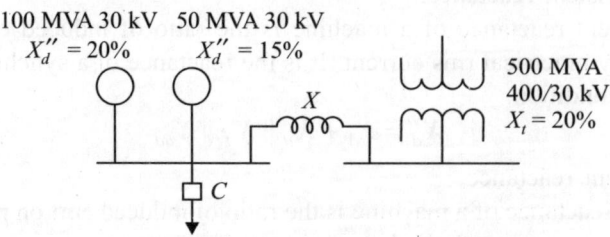

Figure E5.2   One-line diagram for Problem 5.2.

[**Ans.**   1.8 Ω]

5.3 The one-line diagram of a simple power system is shown in Figure E5.3. Each generator is represented by an emf behind the transient reactance. All reactances are expressed in per unit on a common MVA base. The generators are operating on no load at their rated voltage with their emfs in phase. A three-phase fault occurs at bus 1 through a fault impedance of $j0.08$ per unit.

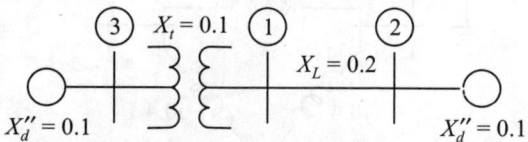

Figure E5.3 One-line diagram for Problem 5.3.

(a) Using Thevenin's theorem, obtain the impedance to the point of fault and the fault current in per unit.

(b) Determine the bus voltages and line currents during the fault.

[**Ans.** $j0.2$, $5\angle{-}90°$, $V_1 = 0.4$, $V_2 = 0.8$, $V_3 = 0.7$ pu]

5.4. The one-line diagram of a simple three-bus power system is shown in Figure E5.4. Each generator is represented by an emf behind the subtransient reactance. All reactances are expressed in per unit on a common MVA base. The generators are operating on no load at their rated voltage. A three-phase fault occurs at bus 3 through a fault reactance of 0.19 per unit.

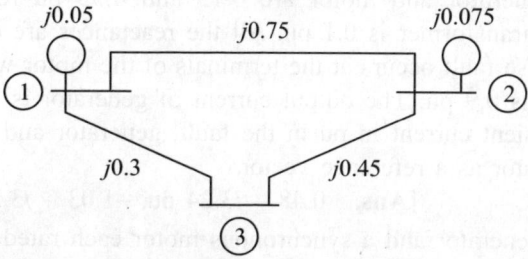

Figure E5.4 One-line diagram for Problem 5.4.

(a) Using Thevenin's theorem obtain the reactance to the point of fault and the fault current in per unit;

(b) Determine the bus voltages and line currents during fault.

[**Ans.** 0.4, $2.5\angle{-}90°$ pu, $V_1 = 0.925$, $V_2 = 0.925$, $V_3 = 0.475$ pu, $I_{12} = 0$, $I_{13} = 1.5\angle{-}90°$, $I_{23} = 1\angle{-}90°$ pu]

5.5 The one-line diagram of a simple three-bus power system is shown in Figure E5.5. Each generator is represented by an emf behind the transient reactance. All reactances are expressed in per unit on a common 100 MVA base. The generators are running on no load at their rated voltage. Determine the fault current, bus voltages and line currents during fault when a balanced three-phase fault with fault impedance $j0.16$ per unit

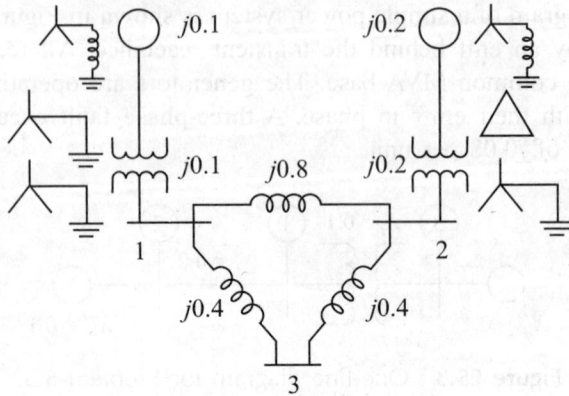

Figure E5.5   One-line diagram for Problem 5.5.

   (i) Occurs on bus 3.
       [**Ans.** $-j2$ pu, 0.76, 0.68, 0.32 pu, $I_{12} = -j0.1$, $I_{13} = -j1.1$, $I_{23} = -j0.9$ pu]
   (ii) Occurs on bus 2.
       [**Ans.** $-j2.5$ pu, 0.8, 0.4, 0.6 pu, $I_{12} = I_{13} = I_{32} = -j0.5$ pu]
   (iii) Occurs on bus 1.
       [**Ans.** $-j3.125$ pu, 0.5, 0.75, 0.625 pu, $I_{21} = I_{31} = I_{23} = -j0.3125$]

5.6 A generator is connected through a transformer to a synchronous motor. The subtransient reactances of generator and motor are 0.15 and 0.35 pu respectively. The leakage reactance of the transformer is 0.1 pu. All the reactances are calculated on a common base. A three-phase fault occurs at the terminals of the motor when the terminal voltage of the generator is 0.9 pu. The output current of generator is 1 pu and 0.8 pf leading. Find the subtransient current in pu in the fault, generator and motor. Use the terminal voltage of generator as a reference vector.
                   [**Ans.** $0.48 - j3.24$ pu, $-1.03 - j3.34$ pu, $-0.55 - j6.58$ pu]

5.7 A synchronous generator and a synchronous motor each rated 25 MVA, 11 kV having 15% subtransient reactance are connected through transformers and a line as shown in Figure E5.6. The transformers are rated 25 MVA, 11/66 kV and 66/11 kV, respectively with leakage reactance of 10% each. The line has a reactance of 10% on a base of 25 MVA, 66 kV. The motor is drawing 15 MW at 0.5 power factor leading and a terminal voltage of 10.6 kV, when a symmetrical three-phase fault occurs at the motor terminals. Find the subtransient current in the generator, motor and fault.

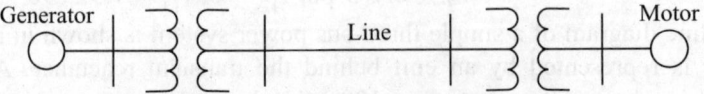

Figure E5.6   Circuit for Problem 5.7.

                   [**Ans.** $3.2795\angle-79.055°$, $5.382\angle-96.6436°$, $-j8.5657$ pu]

5.8 A 60 MVA, Y connected 11 kV synchronous generator is connected to a 60 MVA, 11/132 kV Δ/Y is transformer. The subtransient reactance X_d'' of the generator is

0.12 pu on a 60 MVA base, while the transformer reactance is 0.1 pu on the same base. The generator is unloaded when a symmetrical fault is suddenly placed at point P as shown in Figure E5.7. Find the subtransient symmetrical fault current in pu and actual amperes on both sides of the transformer. Phase to neutral voltage of the generator at no load is 1 pu.

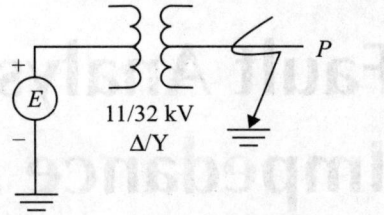

Figure E5.7 System for Problem 5.8.

[**Ans.** $-j4.54$ pu, 14.297 kA, 1.189 kA]

CHAPTER 6

Balanced Fault Analysis using Bus Impedance Matrix

The fault calculation has so far been confined to simple circuits, where network reduction using circuit theory is easy. If it is to be extended to large networks, this method is laborious and an elegant approach is use of bus impedance as a part of Thevenin's method.

6.1 Impedance Matrix Building Algorithm

Bus impedance matrix Z_{bus} is used for short circuit studies while Y_{bus} is used for power flow studies. Z_{bus} may be computed by first forming Y_{bus} and then inverting it, but direct method for building Z_{bus} is very much simpler and straight forward than matrix inversion. This method is also easily amenable for computer solutions. Further any modification of the network does not require complete rebuilding of Z_{bus}.

6.1.1 Modification of Existing Bus Impedance Matrix

There are four cases of modifying an existing Z_{bus}. In our analysis the existing buses will be identified by letters h, i, j and k. The alphabet p will designate a new bus to be added to the network through impedance z_b to convert $n \times n$ matrix Z_{orig} to an $(n + 1) \times (n + 1)$ matrix Z_{new}.

Case 1: Adding z_b from a new bus p to the reference bus

The addition of a new bus p connected to the reference bus through z_b without a connection to any of the buses of the original network cannot alter the original bus voltages when a current I_p is injected at the new bus. Further, the voltage V_p at the new bus equals $I_p z_b$. The modified Z_{bus} given by matrix equation is

$$\begin{pmatrix} V_1 \\ V_2 \\ \vdots \\ V_p \\ \hdashline V_q \end{pmatrix} = \left(\begin{array}{cccc:c} & & & & 0 \\ & Z_{orig} & & & 0 \\ & & & & \vdots \\ & & & & 0 \\ \hdashline 0 & 0 & \cdots & 0 & z_b \end{array} \right) \begin{pmatrix} I_1 \\ I_2 \\ \vdots \\ I_p \\ \hdashline I_q \end{pmatrix} \tag{6.1}$$

Case 2: Adding z_b from a new bus p to an existing bus k

The addition of a new bus p connected through z_b to an existing bus k with current I_p injected at bus p will cause the current entering the original network at bus k to become the sums of I_k which is injected at bus k plus the current I_p coming through z_b as shown in Figure 6.1.

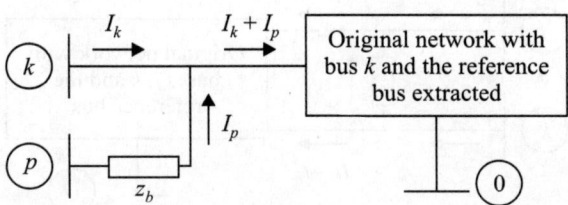

Figure 6.1   Addition of new bus p connected through impedance z_b to existing bus k.

The current I_p flowing into bus k will increase the original V_k by the voltage $I_p Z_{kk}$; that is

$$V_{k(\text{new})} = V_{k(\text{orig})} + I_p Z_{kk} \tag{6.2}$$

and V_p will be larger than the new V_k by the voltage $I_p z_b$

$$V_p = V_{k(\text{orig})} + I_p Z_{kk} + I_p z_b = I_1 Z_{k1} + I_2 Z_{k2} + \cdots + I_n Z_{kn} + I_p(Z_{kk} + z_b) \tag{6.3}$$

The new row to be added to Z_{orig} to find V_p is

$$Z_{k1} \ Z_{k2} \ \cdots \ Z_{kn}(Z_{kk} + z_b) \tag{6.4}$$

Since Z_{bus} must be a square matrix around the principal diagonal we must add a new column which is the transpose of the new row. The new column accounts for the increase of all bus voltages due to I_p. The matrix equation is

$$
\begin{pmatrix} V_1 \\ V_2 \\ \vdots \\ V_n \\ V_q \end{pmatrix}
=
\left(
\begin{array}{cccc:c}
 & & & & Z_{1k} \\
 & Z_{\text{orig}} & & & Z_{2k} \\
 & & & & \vdots \\
 & & & & Z_{nk} \\
\hdashline
Z_{1k} & Z_{2k} & \cdots & Z_{kn} & Z_{kk} + z_b
\end{array}
\right)
\begin{pmatrix} I_1 \\ I_2 \\ \vdots \\ I_n \\ I_q \end{pmatrix}
\tag{6.5}
$$

Note that the n elements of the new row are the elements of row k of Z_{orig} and the n elements of the new column are the elements of column k of Z_{orig}.

Case 3   Adding z_b from an existing bus k to the reference bus

We shall first add a new bus p connected through z_b to bus k and then short-circuit bus p to the reference bus by letting $V_p = 0$ to yield the same matrix equation as Eq. (6.5) except that $V_p = 0$. We then eliminate the $(n + 1)$th row and column using node elimination method (refer Appendix A). Each element Z_{hi} in the new matrix is

$$Z_{hi(\text{new})} = Z_{hi(\text{orig})} - \frac{Z_{h(n+1)} Z_{(n+1)i}}{Z_{kk} + z_b} \tag{6.6}$$

$$Z_{\text{bus(new)}} = Z_{\text{bus(orig)}} - \frac{1}{Z_{kk} + z_b} \begin{pmatrix} Z_{1k} \\ \vdots \\ Z_{nk} \end{pmatrix} \begin{pmatrix} Z_{k1} & \cdots & Z_{kn} \end{pmatrix} \tag{6.7}$$

Case 4 Adding z_b between two existing buses j and k

Figure 6.2 shows the existing buses j and k extracted from the original network. The current I_b is shown flowing through z_b from bus k to bus j.

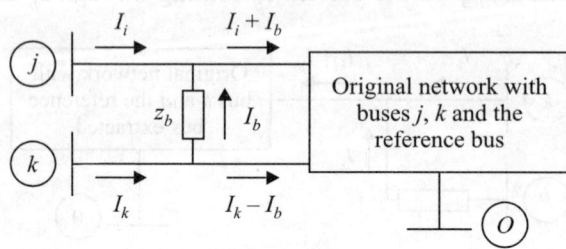

Figure 6.2 Addition of impedance z_b between existing buses j and k.

From Figure 6.2, we get

Node voltage

$$V_1 = Z_{11}I_1 + \cdots + Z_{1j}(I_j + I_b) + Z_{1k}(I_k - I_b) + \cdots$$

By rearranging, we get

$$V_1 = Z_{11}I_1 + \cdots + Z_{1j}I_j + Z_{1k}I_k + \cdots + I_b(Z_{1j} - Z_{1k})$$

Similarly,

$$V_j = Z_{j1}I_1 + \cdots + Z_{jj}I_j + Z_{jk}I_k + \cdots + I_b(Z_{jj} - Z_{jk}) \tag{6.8}$$

and

$$V_k = Z_{k1}I_1 + \cdots + Z_{kj}I_j + Z_{kk}I_k + \cdots + I_b(Z_{kj} - Z_{kk})$$

We need one more equation since I_b is unknown. From Figure 6.2,

$$V_k - V_j = I_b z_b \tag{6.9}$$

$$0 = I_b z_b + V_j - V_k$$

By substituting for V_j and V_k and using the relation $Z_{jk} = Z_{kj}$, we obtain

$$0 = I_b z_b + (Z_{j1} - Z_{k1})I_1 + \cdots + (Z_{jj} - Z_{kj})I_j + \cdots (Z_{jk} - Z_{kk})I_k + \cdots + (Z_{jj} + Z_{kk} - 2Z_{jk})I_b$$
$$\text{since } Z_{kj} = Z_{jk}$$

Collecting the coefficients of I_b and naming their sum as Z_{bb}

$$Z_{bb} = Z_{jj} + Z_{kk} - 2Z_{jk} + z_b \tag{6.10}$$

The matrix equation is

$$
\begin{pmatrix} V_1 \\ \vdots \\ V_j \\ \vdots \\ V_k \\ \vdots \\ V_n \\ 0 \end{pmatrix}
=
\left(
\begin{array}{cccc:c}
 & & & & Z_{1j} - Z_{1k} \\
 & & & & \vdots \\
 & & Z_{\text{orig}} & & Z_{jj} - Z_{jk} \\
 & & & & \vdots \\
 & & & & Z_{kj} - Z_{kk} \\
 & & & & \vdots \\
 & & & & Z_{nj} - Z_{nk} \\
\hdashline
Z_{j1} - Z_{k1} & \cdots & Z_{jj} - Z_{kj} & \cdots & Z_{jk} - Z_{kk} \quad \cdots \quad Z_{jn} - Z_{kn} & Z_{bb}
\end{array}
\right)
\begin{pmatrix} I_1 \\ \vdots \\ I_j \\ \vdots \\ I_k \\ \vdots \\ I_n \\ I_b \end{pmatrix}
\tag{6.11}
$$

The new column is column j minus column k of Z_{orig} with Z_{bb} in the $(n + 1)$th row. The new row is the transpose of the column. We then eliminate the $(n + 1)$th row and column using node elimination method. Each element Z_{hi} in the new matrix is

$$Z_{hi(\text{new})} = Z_{hi(\text{orig})} - \frac{Z_{h(n+1)}Z_{(n+1)i}}{Z_{jj} + Z_{kk} - 2Z_{jk} + z_b} \tag{6.12}$$

$$Z_{\text{bus(new)}} = Z_{\text{bus(orig)}} - \frac{1}{Z_{jj} + Z_{kk} - 2Z_{jk} + z_b} \begin{pmatrix} (Z_{1j} - Z_{1k}) \\ \vdots \\ (Z_{nj} - Z_{nk}) \end{pmatrix} ((Z_{j1} - Z_{k1}) \cdots (Z_{jn} - Z_{kn})) \tag{6.13}$$

Note Case involving two new buses can be considered one after another.

6.1.2 Direct Determination of Bus Impedance Matrix

We start by writing the equation for one bus connected through an impedance z_b to the reference bus as

$$V_1 = z_b I_1 \tag{6.14}$$

This can be considered as a matrix equation which has one row and one column. We use the bus impedance modification technique to add new buses. Usually, the buses of the network are numbered in the order in which they are added.

Algorithm for building Z_{bus} matrix

Step 1   Get the bus and line data.

Step 2   Introduce buses connected to reference bus by connecting lines one by one between them and reference bus. (Impedance of connecting line z_b is diagonal and other elements are 0)

$$Z_{\text{ref}} = \begin{pmatrix} z_{b1} & 0 & 0 \\ 0 & z_{b2} & 0 \end{pmatrix}$$

Step 3   Introduce the other buses connected to existing buses by connecting lines one by one between them and existing buses k. (Impedance of connecting line is z_b) as given below.

$$Z_{\text{exist}} = \begin{pmatrix} & & & & Z_{1k} \\ & & & & Z_{2k} \\ & & Z_{\text{ref}} & & \vdots \\ & & & & Z_{nk} \\ Z_{k1} & Z_{k2} & \cdots & \cdots & Z_{kn} & Z_{kk} + z_b \end{pmatrix}$$

Step 4   Add links between existing buses j and k one by one as shown below (Impedance of connecting link is z_b) and reduce the matrix to original size after every addition.

$$\begin{pmatrix} & & & & \vdots & Z_{1j} - Z_{1k} \\ & & & & & \vdots \\ & & Z_{\text{exist}} & & & Z_{jj} - Z_{jk} \\ & & & & & \vdots \\ & & & & & Z_{kj} - Z_{kk} \\ & & & & & \vdots \\ & & & & & Z_{nj} - Z_{nk} \\ \hline (Z_{j1} - Z_{k1}) & \cdots & (Z_{jj} - Z_{kj}) & \cdots & (Z_{jk} - Z_{kk}) \cdots (Z_{jn} - Z_{kn}) & Z_{bb} \end{pmatrix}$$

where $Z_{bb} = Z_{jj} + Z_{kk} - 2Z_{jk} + z_b$

$$Z_{\text{bus}} = Z_{\text{exist}} - \frac{1}{Z_{ij} + Z_{kk} - 2Z_{jk} + z_b} \begin{pmatrix} (Z_{1j} - Z_{1k}) \\ \vdots \\ (Z_{nj} - Z_{nk}) \end{pmatrix} [(Z_{j1} - Z_{k1}) \cdots (Z_{jn} - Z_{kn})]$$

EXAMPLE 6.1 Determine Z_{bus} for the network shown in Figure 6.3(a) whose impedances are given in per unit.

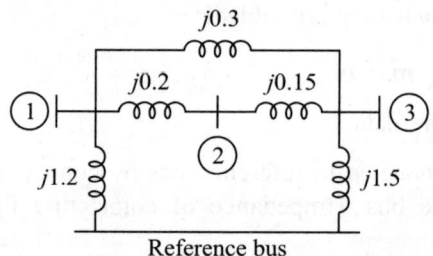

Figure 6.3(a) Network for Example 6.1.

Solution For ease of computation let us interchange the numbering of the buses 2 and 3. The modified network is shown in Figure 6.3(b).

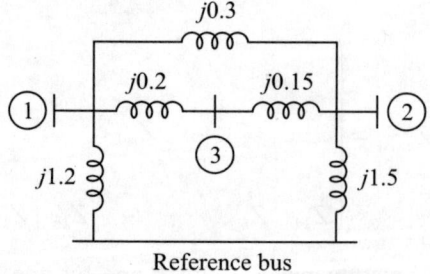

Figure 6.3(b) Network for Example 6.1.

We start by establishing buses 1 and 2 with their impedances to the reference bus.

$$Z_{\text{bus}} = j\begin{pmatrix} 1.2 & 0 \\ 0 & 1.5 \end{pmatrix}$$

Next we establish bus 3 with its impedance $j0.2$ to bus 1. The new bus impedance matrix is

$$j\begin{pmatrix} 1.2 & 0 & 1.2 \\ 0 & 1.5 & 0 \\ 1.2 & 0 & 1.4 \end{pmatrix}$$

Since bus 1 is the bus to which bus 3 is being connected, the term $j1.4(Z_{33})$ is the sum of Z_{11} of the matrix being modified and the impedance z_b of the branch being connected to bus 1 from bus 3. The other elements of the new row and column are the repetition of the row 1 and column 1 of the matrix being modified, because the new bus is being connected to bus 1.

To add the impedance $z_b = j0.3$ from bus 2 to bus 1, a new bus 4 is created to obtain the new impedance matrix.

$$j\begin{pmatrix} 1.2 & 0 & 1.2 & 1.2 \\ 0 & 1.5 & 0 & -1.5 \\ 1.2 & 0 & 1.4 & 1.2 \\ \hline 1.2 & -1.5 & 1.2 & 3 \end{pmatrix}$$

The new elements of row 4 and column 4 are

$$Z_{14} = Z_{11} - Z_{12} = j1.2 = Z_{41}$$
$$Z_{24} = Z_{21} - Z_{22} = -j1.5 = Z_{42}$$
$$Z_{34} = Z_{31} - Z_{32} = j1.2 = Z_{43}$$
$$Z_{44} = Z_{11} + Z_{22} - 2Z_{12} + z_b = j3$$

The row 4 and column 4 are eliminated by node elimination method.

$$Z_{11} = j\left(1.2 - \frac{1.2 \times 1.2}{3}\right) = j0.72$$

$$Z_{23} = Z_{32} = j\left(0 - \frac{1.2 \times (-1.5)}{3}\right) = j0.6$$

Other elements of the matrix are calculated similarly. The reduced matrix is

$$j\begin{pmatrix} 0.72 & 0.60 & 0.72 \\ 0.60 & 0.75 & 0.60 \\ 0.72 & 0.60 & 0.92 \end{pmatrix}$$

Finally, we add the impedance $z_b = j0.15$ from bus 2 to bus 3. A new bus 4 is created to obtain the new impedance matrix.

$$j\begin{pmatrix} 0.72 & 0.60 & 0.72 & -0.12 \\ 0.60 & 0.75 & 0.60 & 0.15 \\ 0.72 & 0.60 & 0.92 & -0.32 \\ \hline -0.12 & 0.15 & -0.32 & 0.62 \end{pmatrix}$$

The new elements of row 4 and column 4 are

$$Z_{14} = Z_{12} - Z_{13} = -j0.12 = Z_{41}$$
$$Z_{24} = Z_{22} - Z_{23} = j0.15 = Z_{42}$$
$$Z_{34} = Z_{32} - Z_{33} = -j0.32 = Z_{43}$$
$$Z_{44} = Z_{22} + Z_{33} - 2Z_{23} + z_b = j0.62$$

The row 4 and column 4 are eliminated by node elimination method.

$$Z_{11} = j\left(0.72 - \frac{(-0.12) \times (-0.12)}{0.92}\right) = j0.6968$$

$$Z_{23} = Z_{32} = j\left(0.60 - \frac{(-0.32) \times (-0.12)}{0.62}\right) = j0.6774$$

Other elements of the matrix are calculated similarly. The reduced matrix is

$$j\begin{pmatrix} 0.6968 & 0.6290 & 0.6581 \\ 0.6290 & 0.7137 & 0.6774 \\ 0.6581 & 0.6774 & 0.7548 \end{pmatrix}$$

We revert back to original network by interchanging rows and columns corresponding to buses 2 and 3. Interchanging rows 2 and 3, we get

$$j\begin{pmatrix} 0.6968 & 0.6290 & 0.6581 \\ 0.6581 & 0.6774 & 0.7548 \\ 0.6290 & 0.7137 & 0.6774 \end{pmatrix}$$

Now we interchange columns 2 and 3 to get final Z_{bus} as

$$j\begin{pmatrix} 0.6968 & 0.6581 & 0.6290 \\ 0.6581 & 0.7548 & 0.6774 \\ 0.6290 & 0.6774 & 0.7137 \end{pmatrix}$$

EXAMPLE 6.2   The bus impedance matrix of a network is given below.

$$Z_{bus} = j\begin{pmatrix} 0.183 & 0.078 & 0.141 \\ 0.078 & 0.148 & 0.106 \\ 0.141 & 0.106 & 0.267 \end{pmatrix}$$

One of the lines between buses 1 and 3 with impedance $j0.56$ is removed by the simultaneous opening of breakers at both ends of the line. Determine the new bus impedance matrix.

Solution   The removal of an element is equivalent to connecting a link having impedance equal to the negated value of the original impedance. So link $z_{13} = -j0.56$ between node $q = 3$ and node $p = 1$ is added.

$$Z_{bus} = \begin{pmatrix} Z_{11} & Z_{12} & Z_{13} & Z_{13} - Z_{11} \\ Z_{21} & Z_{22} & Z_{23} & Z_{23} - Z_{21} \\ Z_{31} & Z_{32} & Z_{33} & Z_{33} - Z_{31} \\ Z_{31} - Z_{11} & Z_{32} - Z_{12} & Z_{33} - Z_{13} & Z_{44} \end{pmatrix}$$

where $Z_{44} = Z_{11} + Z_{33} - 2Z_{13} + z_{13}$

$$= j \begin{pmatrix} 0.183 & 0.078 & 0.141 & -0.042 \\ 0.078 & 0.148 & 0.106 & 0.028 \\ 0.141 & 0.106 & 0.267 & 0.126 \\ -0.042 & 0.028 & 0.126 & -0.392 \end{pmatrix}$$

$$\frac{\Delta Z (\Delta Z)^T}{Z_{44}} = j \frac{1}{-0.392} \begin{pmatrix} -0.042 \\ 0.028 \\ 0.126 \end{pmatrix} (-0.042 \quad 0.028 \quad 0.126)$$

$$= j \begin{pmatrix} -0.0045 & 0.003 & 0.0135 \\ 0.003 & -0.002 & -0.009 \\ 0.0135 & -0.009 & -0.0405 \end{pmatrix}$$

New bus impedance matrix:

$$= j \begin{pmatrix} 0.183 & 0.078 & 0.141 \\ 0.078 & 0.148 & 0.106 \\ 0.141 & 0.106 & 0.267 \end{pmatrix} - j \begin{pmatrix} -0.0045 & 0.003 & 0.0135 \\ 0.003 & -0.002 & -0.009 \\ 0.0135 & -0.009 & -0.0405 \end{pmatrix}$$

$$= j \begin{pmatrix} 0.1875 & 0.075 & 0.1275 \\ 0.075 & 0.15 & 0.115 \\ 0.1275 & 0.115 & 0.3075 \end{pmatrix}$$

6.2 Algorithm for Short-circuit/Fault Studies

Consider an n-bus system shown in Figure 6.4 having a steady load. The following steps are followed to find the fault current:

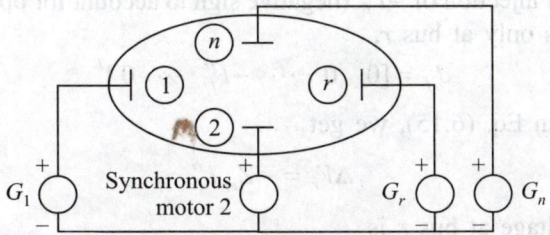

Figure 6.4 Sample n-bus system.

Step 1   Calculate pre-fault voltages at all buses and currents in all transmission lines and transformers by conducting power flow study. Let the pre-fault voltages be

$$V^0_{bus} = [V^0_1 \quad V^0_2 \quad \cdots \quad V^0_r \quad \cdots \quad V^0_n]^T$$

Let us assume that a fault occurs at rth bus through a fault impedance z_f. The post-fault voltage is given by

$$V^f_{bus} = V^0_{bus} + \Delta V_{bus}$$

where ΔV_{bus} is the change in bus voltage caused by fault.

Step 2   Draw the passive Thevenin's network as shown in Figure 6.5 for the system with synchronous machines replaced by either transient or subtransient reactances depending on the type of short circuit current to be computed-transient or subtransient.

Step 3   Excite the Thevenin's network with $-V_r^0$ in series with z_f as shown in Figure 6.5 to simulate the fault conditions.

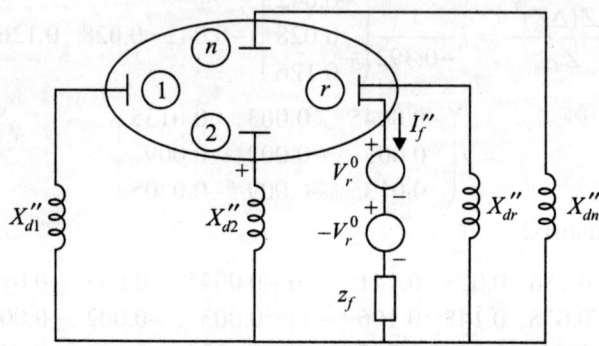

Figure 6.5   Passive Thevenin's network with fault simulation at bus r.

It may be noted that V_r^0 alone does not cause any current in the branch. Changes in voltages are given by

$$\Delta V_{\text{bus}} = Z_{\text{bus}} J_f \tag{6.15}$$

where

Z_{bus} = bus impedance matrix of Thevenin's network

J_f = bus current injection vector.

As the current injection of $-I_f''$ (negative sign to account for opposite direction shown in Figure 6.5) is only at bus r,

$$J_f = [0 \quad 0 \quad \cdots \quad -I_f'' \quad \cdots \quad 0]^T \tag{6.16}$$

Substituting in Eq. (6.15), we get

$$\Delta V_r = -Z_{rr} I_f'' \tag{6.17}$$

Step 4   Post-fault voltage at bus r is

$$V_r^f = V_r^0 + \Delta V_r = V_r^0 - Z_{rr} I_f'' \tag{6.18}$$

We know

$$V_r^f = z_f I_f'' \tag{6.19}$$

$$\therefore \qquad z_f I_f'' = V_r^0 - Z_{rr} I_f''$$

$$I_f'' = \frac{V_r^0}{Z_{rr} + z_f} \tag{6.20}$$

At ith bus change in voltage

$$\Delta V_i = -Z_{ir} I_f'' \tag{6.21}$$

$$V_i^f = V_i^0 + \Delta V_i = V_i^0 - Z_{ir}I_f''$$

$$= V_i^0 - \frac{Z_{ir}}{Z_{rr} + z_f} V_r^0 \quad i = 1, 2, ..., N \tag{6.22}$$

using Eq. (6.20)

Fault contribution by lines/transformers $(i - j)$

$$I_{ijf}'' = \frac{V_i^f - V_j^f}{z_{ij}} \tag{6.23}$$

where z_{ij} is the impedance of the line between buses i and j and not the element Z_{ij} of the bus impedance matrix.

This current is caused by fault only. Total fault current is obtained by superposing this current over the pre-fault current caused by all generators with transient or subtransient reactances as the case may be

$$I_{ij}'' = I_{ijf}'' + I_{ij} \tag{6.24}$$

Flow chart for calculating symmetrical fault current using Z_{bus} is given in Figure 6.6.

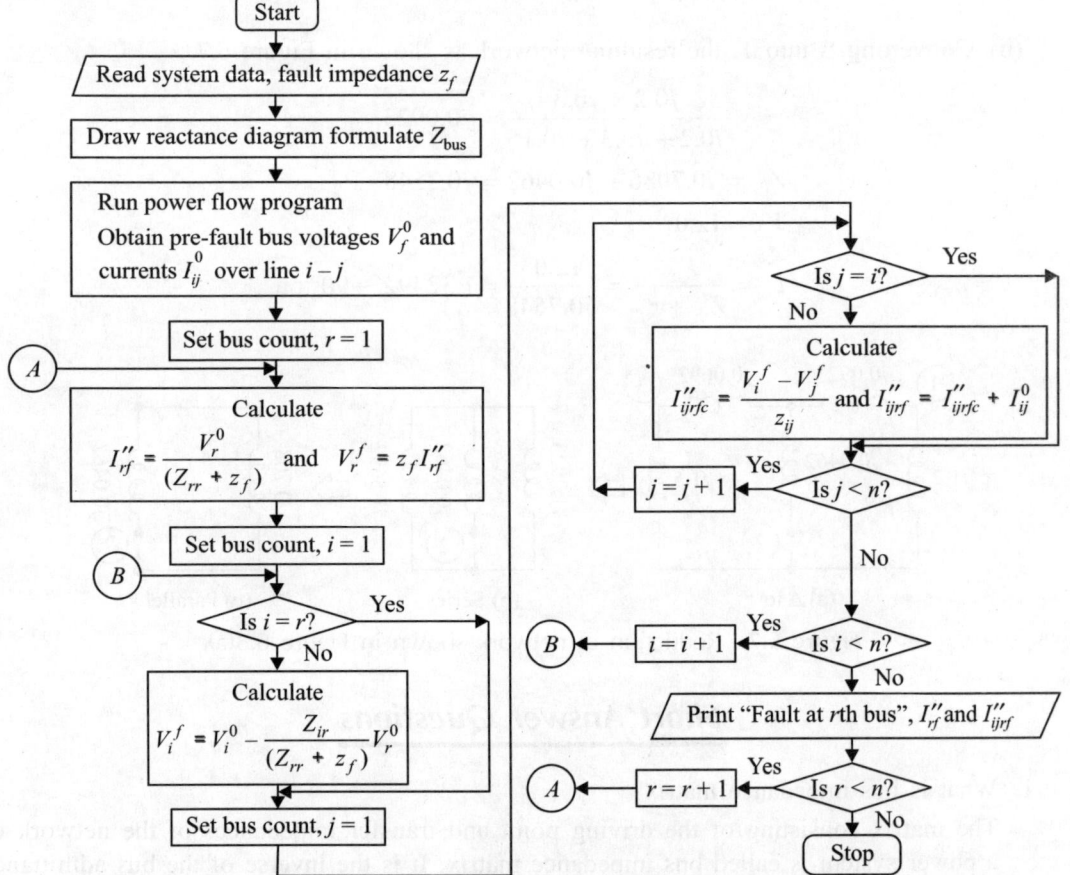

Figure 6.6 Flow chart for calculation of symmetrical fault current using Z_{bus}.

EXAMPLE 6.3 Find the fault current for a three-phase solid/bolted short circuit at bus 2 in the network shown in Figure 6.3(a) using (i) Z_{bus} (ii) network reduction.

Solution

(i) Fault with zero fault impedance is called solid or bolted fault.

It may be noted that the diagonal elements of the Z_{bus} are Thevenin impedance seen from the bus, since when the current injections at all buses except the bus under consideration is made zero, only diagonal element will cause the voltage drop as shown below.

$$V_j = Z_{j1}I_1 + \cdots Z_{jj}I_j + \cdots Z_{jn}I_n \tag{6.25}$$

$$Z_{jj} = \frac{V_j}{I_j}\bigg|_{I_i=0 \text{ for } i=1,\dots n \neq j} \tag{6.26}$$

From Z_{bus} calculated from Example 6.1, $Z_{22} = j0.7548$ pu

$$I_f = \frac{V_2^0}{Z_{22} + z_f} = \frac{1\angle 0°}{j0.7548} = 1.3249\angle -90° \text{ pu}$$

(ii) Converting Δ into Y, the resulting network is shown in Figure 6.7.

$$\frac{j0.2 \times j0.3}{j0.2 + j0.3 + j0.15} = j0.0923$$

$$Z_{th} = j0.7086 + j0.0462 = j0.7548$$

$$V_{th} = 1\angle 0°$$

$$I_f = \frac{V_{th}}{Z_{th} + z_f} = \frac{1\angle 0°}{j0.7548} = 1.3249\angle -90° \text{ pu}$$

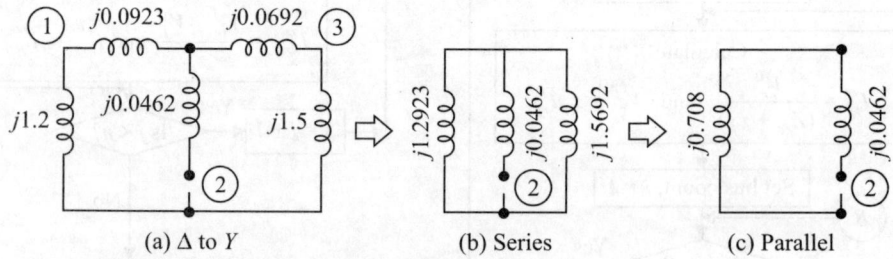

(a) Δ to Y (b) Series   (c) Parallel

Figure 6.7 Reduction of network shown in Figure 6.3(a).

Short Answer Questions

1. What is bus impedance matrix?

The matrix consisting of the driving point and transfer impedances of the network of a power system is called bus impedance matrix. It is the inverse of the bus admittance matrix and is symmetrical.

2. Name the diagonal elements and off diagonal elements of bus impedance matrix.
   The diagonal elements of bus impedance matrix are called driving point impedances of the buses and off diagonal elements of bus impedance matrix are called transfer impedances of the buses.

3. What are the methods available for forming bus impedance matrix (Z_{bus}).
   1. Z_{bus} can be computed by first forming Y_{bus} and then inverting it
   2. Direct method which utilizes the techniques of modifications of an existing Z_{bus}.

4. What are the four cases of modifying an existing Z_{bus}?
   Case 1:   Adding z_b from a new bus p to the reference bus.
   Case 2:   Adding z_b from a new bus p to an existing bus k.
   Case 3:   Adding z_b from an existing bus k to the reference bus.
   Case 4:   Adding z_b between two existing buses j and k.

Exercises

6.1 Using the method of building algorithm, find the bus impedance matrix for the network shown in Figure E6.1.

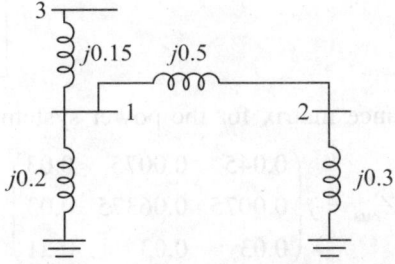

Figure E6.1   Network for Problem 6.1.

$$\mathbf{Ans.}\begin{bmatrix} j0.24 & j0.14 & j0.2 & j0.14 \\ j0.14 & j0.2275 & j0.175 & j0.2275 \\ j0.2 & j0.175 & j0.31 & j0.175 \\ j0.14 & j0.2275 & j0.175 & j0.4175 \end{bmatrix}$$

6.2 Obtain the bus impedance matrix for the network of Problem 5.3.

$$\mathbf{Ans.}\begin{bmatrix} j0.12 & j0.04 & j0.06 \\ j0.04 & j0.08 & j0.02 \\ j0.06 & j0.02 & j0.08 \end{bmatrix}$$

6.3 Obtain the bus impedance matrix for the network of Problem 5.4.

$$\mathbf{Ans.}\begin{bmatrix} j0.045 & j0.0075 & j0.03 \\ j0.0075 & j0.06375 & j0.03 \\ j0.03 & j0.03 & j0.21 \end{bmatrix}$$

6.4 The bus impedance matrix for the network shown in Figure E6.2 is

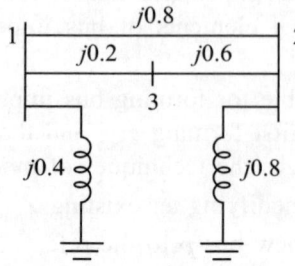

Figure E6.2   One-line diagram for Problem 6.4.

$$Z_{\text{bus}} = j \begin{pmatrix} 0.3 & 0.2 & 0.275 \\ 0.2 & 0.4 & 0.25 \\ 0.275 & 0.25 & 0.41875 \end{pmatrix}$$

There is a line outage and the line from bus 1 to 2 is removed. Using the method of building algorithm determine the new bus impedance matrix from the original one.

$$\left[\textbf{Ans.} \quad j \begin{pmatrix} 0.32 & 0.16 & 0.28 \\ 0.16 & 0.48 & 0.24 \\ 0.28 & 0.24 & 0.42 \end{pmatrix} \right]$$

6.5 The per unit bus impedance matrix for the power system of Problem 5.4 is given by

$$Z_{\text{bus}} = j \begin{pmatrix} 0.045 & 0.0075 & 0.03 \\ 0.0075 & 0.06375 & 0.03 \\ 0.03 & 0.03 & 0.21 \end{pmatrix}$$

A three-phase fault occurs at bus 3 through a fault impedance of $j0.19$ per unit. Using the bus impedance matrix calculate the fault current, bus voltages and line currents during fault.   [**Ans.**   Same as Problem 5.4]

6.6 Determine the fault current, bus voltages and line currents during fault for Problem 5.5 using bus impedance method.

[**Ans.**   (i) $-j2$ pu, 0.76, 0.68, 0.32 pu, $I_{12} = -j0.1$, $I_{13} = -j1.1$, $I_{23} = -j0.9$ pu;

(ii) $-j2.5$ pu, 0.8, 0.4, 0.6 pu, $I_{12} = I_{13} = I_{32} = -j0.5$ pu;

(iii) $-j3.125$ pu, 0.5, 0.75, 0.625 pu, $I_{21} = I_{31} = I_{23} = -j0.3125$]

6.7 Find the bus impedance matrix for the system whose reactance diagram is shown in Figure E6.3. All the reactances are in pu.

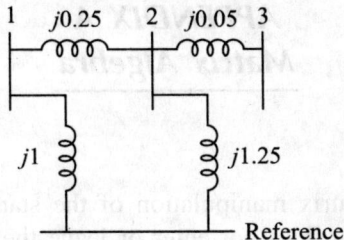

Figure E6.3 Reactance diagram for Problem 6.7.

$$\text{Ans.} \begin{bmatrix} j0.6 & j0.5 & j0.5 \\ j0.5 & j0.625 & j0.625 \\ j0.5 & j0.625 & j0.675 \end{bmatrix}$$

6.8 Find the bus impedance matrix for the system whose reactance diagram is shown in Figure E6.4. All the impedances are in pu.

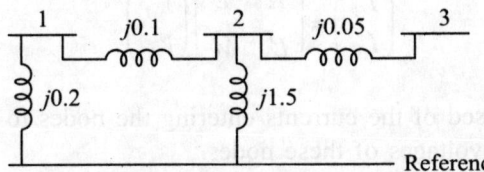

Figure E6.4 Reactance diagram for Problem 6.8.

$$\text{Ans.} \begin{bmatrix} j0.177 & j0.166 & j0.166 \\ j0.166 & j0.25 & j0.25 \\ j0.166 & j0.25 & j0.3 \end{bmatrix}$$

APPENDIX A
Matrix Algebra

A.1 Node Elimination

Nodes may be eliminated by matrix manipulation of the standard node equations. However, only those nodes at which current does not enter or leave the network can be eliminated.

The standard node equations in matrix form are

$$I = Y_{\text{bus}}V \tag{A.1}$$

where I and V are column matrices or simply vectors and Y_{bus} is a symmetrical square matrix.

The column matrices must be so arranged that elements associated with nodes to be eliminated are in the lower rows of the matrices. Elements of the square matrix are located correspondingly. The matrices are partitioned so that the elements associated with the nodes to be eliminated are separated from the other elements.

$$\begin{pmatrix} I_A \\ I_X \end{pmatrix} = \begin{pmatrix} K & L \\ L^T & M \end{pmatrix} \begin{pmatrix} V_A \\ V_X \end{pmatrix} \tag{A.2}$$

where

I_X = submatrix composed of the currents entering the nodes to be eliminated.
V_X = submatrix of the voltages of these nodes.
K = square submatrix, the elements of which are self and mutual admittances associated with nodes to be retained.
M = square submatrix, the elements of which are self and mutual admittances associated with nodes to be eliminated.
L and its transpose L^T = submatrices composed of only those mutual admittances common to nodes to be retained and to those to be eliminated.

Performing multiplication

$$I_A = KV_A + LV_X \tag{A.3}$$
$$I_X = L^TV_A + MV_X \tag{A.4}$$

Since all the elements of I_X are zero

$$MV_X = -L^TV_A$$
$$V_X = -M^{-1}L^TV_A$$

Substituting this in Eq. (A.3)

$$I_A = KV_A - LM^{-1}L^TV_A = (K - LM^{-1}L^T)V_A$$
$$\therefore \qquad Y_{\text{bus}} = K - LM^{-1}L^T \tag{A.5}$$

EXAMPLE A.1 Eliminate the nodes 3 and 4 from the admittance matrix given below:

$$\begin{pmatrix} -j9.8 & 0 & j4 & j5 \\ 0 & -j8.3 & j2.5 & j5 \\ j4 & j2.5 & -j14.5 & j8 \\ j5 & j5 & j8 & -j18 \end{pmatrix}$$

Solution Partitioning for eliminating nodes 3 and 4 from 1 and 2.

$$K = \begin{pmatrix} -j9.8 & 0 \\ 0 & -j8.3 \end{pmatrix} \quad L = \begin{pmatrix} j4 & j5 \\ j2.5 & j5 \end{pmatrix} \quad M = \begin{pmatrix} -j14.5 & j8 \\ j8 & -j18 \end{pmatrix}$$

$$M^{-1} = \frac{1}{-197} \begin{pmatrix} -j18 & -j8 \\ -j8 & -j14.5 \end{pmatrix} = \begin{pmatrix} j0.0914 & j0.0406 \\ j0.0406 & j0.0736 \end{pmatrix}$$

$$LM^{-1}L^T = \begin{pmatrix} j4 & j5 \\ j2.5 & j5 \end{pmatrix} \begin{pmatrix} j0.0914 & j0.0406 \\ j0.0406 & j0.0736 \end{pmatrix} \begin{pmatrix} j4 & j2.5 \\ j5 & j5 \end{pmatrix}$$

$$= -\begin{pmatrix} j4.9264 & j4.0736 \\ j4.0736 & j3.4264 \end{pmatrix}$$

$$Y_{\text{bus}} = K - LM^{-1}L^T = \begin{pmatrix} -j9.8 & 0 \\ 0 & -j8.3 \end{pmatrix} + \begin{pmatrix} j4.9264 & j4.0736 \\ j4.0736 & j3.4264 \end{pmatrix}$$

$$= \begin{pmatrix} -j4.8736 & j4.0736 \\ j4.0736 & -j4.8736 \end{pmatrix}$$

A.2 Node Elimination Node by Node

The matrix partitioning method is a general method more suitable for computer solutions unlike Y–Δ transformations and series parallel combinations. However, for the elimination of a large number of nodes, the elements of matrix M whose inverse must be found will be large.

Inverting a matrix is avoided by eliminating one node at a time and the process is very simple and repetitive and suitable for computer solutions. The node to be eliminated must be the highest numbered node and renumbering may be required. The matrix M becomes a single element and M^{-1} is the reciprocal of the element. The original admittance matrix is partitioned into submatrices.

$$Y_{\text{bus}} = \left. \begin{pmatrix} Y_{11} & \cdots & Y_{1i} & \cdots & Y_{1n} \\ \vdots & & & & \\ Y_{h1} & \cdots & \boxed{Y_{hi}} & \cdots & \boxed{Y_{hn}} \\ \vdots & & & & \\ Y_{n1} & \cdots & \boxed{Y_{ni}} & \cdots & Y_{nn} \end{pmatrix} \right\} L \tag{A.6}$$

The reduced $(n-1) \times (n-1)$ matrix according to Eq. (A.5) will be

$$Y_{\text{bus}} = \begin{pmatrix} Y_{11} & \cdots & Y_{1i} & \cdots \\ \vdots & & & \\ Y_{h1} & \cdots & Y_{hi} & \cdots \\ \vdots & & & \end{pmatrix} - \frac{1}{Y_{nn}} \begin{pmatrix} Y_{1n} \\ \vdots \\ Y_{hn} \\ \vdots \end{pmatrix} (Y_{n1} \cdots Y_{ni} \cdots) \tag{A.7}$$

When the indicated manipulation is accomplished, the element in row h and column i of the resulting $(n - 1) \times (n - 1)$ matrix will be

$$Y_{hi(\text{new})} = Y_{hi(\text{orig})} - \frac{Y_{hn}Y_{ni}}{Y_{nn}} \tag{A.8}$$

That is to modify the element Y_{hi} (enclosed by circle) in row h column i by subtracting from it the product of the elements at the last row and column (enclosed by rectangle) corresponding to the element to be modified divided by the element in the lowest right corner.

EXAMPLE A.2   Perform the node elimination of Example A.1 first by removing node 4 and then node 3.

Solution   The original matrix is partitioned for removal of one node 4.

$$Y_{\text{bus}} = \begin{pmatrix} -j9.8 & 0 & j4 & j5 \\ 0 & -j8.3 & j2.5 & j5 \\ j4 & j2.5 & -j14.5 & j8 \\ j5 & j5 & j8 & -j18 \end{pmatrix}$$

To modify the element Y_{32} in row 3 column 2 subtract from it the product of the elements enclosed by rectangles and divide by the element in the lowest right corner.

$$Y_{hi(\text{new})} = Y_{hi(\text{orig})} - \frac{Y_{hn}Y_{ni}}{Y_{nn}}$$

$$Y_{32} = j2.5 - \frac{j8 \times j5}{-j18} = j4.7222$$

Similarly,

$$Y_{11} = -j9.8 - \frac{j5 \times j5}{-j18} = -j8.4111$$

Other elements are found in the same manner to yield

$$Y_{\text{bus}} = \begin{pmatrix} -j8.4111 & j1.3889 & j6.2222 \\ j1.3889 & -j6.9111 & j4.7222 \\ j6.2222 & j4.7222 & -j10.9444 \end{pmatrix}$$

Reducing the above matrix to remove node 3 yields:

$$Y_{\text{bus}} = \begin{pmatrix} -j4.8736 & j4.0736 \\ j4.0736 & -j4.8736 \end{pmatrix}$$

which is identical to the matrix found by the matrix partitioning method where two nodes were removed simultaneously.

CHAPTER 7

Symmetrical Components and Sequence Networks

For a balanced system the analysis is made on a per phase basis. When the system is unbalanced, the voltage, current and the impedance of the different phases become unequal. Such a system can be resolved into a balanced system using a method of symmetrical components. Thus, per phase analysis can be extended to unbalanced system also.

7.1 Symmetrical Components

7.1.1 Fortescue's Theorem

Fortescue's theorem states that an unbalanced system of n related phasors can be resolved into n balanced phasors called symmetrical components of the original unbalanced phasors.

For a three-phase system, the three unbalanced phasors can be resolved into three balanced phasors. The balanced or symmetrical components are:

- Positive-sequence components consisting of three phasors equal in magnitude, displaced from each other by 120° in phase, and having the same phase sequence as the original phasors.
- Negative-sequence components consisting of three phasors equal in magnitude, displaced from each other by 120° in phase, and having the phase sequence opposite to that of the original phasors.
- Zero-sequence components consisting of three phasors equal in magnitude and with zero phase displacement from each other.

7.1.2 Phase Sequence

Phase sequence of the phasors is the order in which they pass through a positive maximum when they rotate in anticlockwise direction. The phase sequence *abc* implies that the maxima of the three phasors occur in the order a followed by b then by c. If abc is taken as positive phase sequence then acb represents the negative phase sequence, as the rotation is in clockwise direction.

7.1.3 Operator a

Similar to the operator j whose magnitude is unity and phase angle is 90°, operator a has a magnitude of unity and phase angle of 120°. If this operator is applied to a phasor, the phasor is rotated through 120° in anticlockwise direction. Successive application rotates through 120° for each application.

$$a = 1\angle 120° = 1e^{j2\pi/3} = -0.5 + j0.866 \tag{7.1}$$
$$a^2 = 1\angle 240° = -0.5 - j0.866$$
$$a^3 = 1\angle 360° = 1$$

It may be noted that $1 + a + a^2 = 0$

If V_a, V_b and V_c represent a set of unbalanced voltage phasors, the three sets of balanced phasors are

Positive-sequence components	V_{a1}, V_{b1}, V_{c1}
Negative-sequence components	V_{a2}, V_{b2}, V_{c2}
Zero-sequence components	V_{a0}, V_{b0}, V_{c0}

The symmetrical components are shown in Figure 7.1

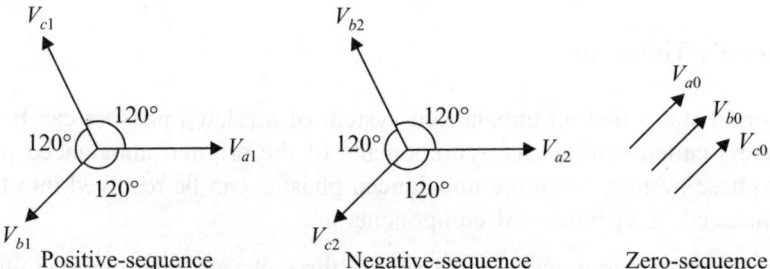

Figure 7.1 Symmetrical components.

The subscripts 0, 1 and 2 refer to zero-sequence, positive-sequence and negative-sequence, respectively.

7.1.4 Voltage and Current in Terms of Symmetrical Components

The phase voltages V_a, V_b and V_c can be written as the sum of their respective symmetrical components.

$$\left. \begin{array}{l} V_a = V_{a0} + V_{a1} + V_{a2} \\ V_b = V_{b0} + V_{b1} + V_{b2} \\ V_c = V_{c0} + V_{c1} + V_{c2} \end{array} \right\} \tag{7.2}$$

Expressing the symmetrical component quantities of b and c phases in terms of a phase using operator a, we get

$$V_a = V_{a0} + V_{a1} + V_{a2}$$
$$V_b = V_{a0} + a^2 V_{a1} + a V_{a2}$$
$$V_c = V_{a0} + a V_{a1} + a^2 V_{a2}$$

In matrix form

$$\begin{pmatrix} V_a \\ V_b \\ V_c \end{pmatrix} = \begin{pmatrix} 1 & 1 & 1 \\ 1 & a^2 & a \\ 1 & a & a^2 \end{pmatrix} \begin{pmatrix} V_{a0} \\ V_{a1} \\ V_{a2} \end{pmatrix} \tag{7.3}$$

i.e.

$$V^{abc} = A V^{012} \tag{7.4}$$

where

$$A = \begin{pmatrix} 1 & 1 & 1 \\ 1 & a^2 & a \\ 1 & a & a^2 \end{pmatrix}$$

is called the symmetrical component transformation matrix.

From Eq. (7.4)

$$V^{012} = A^{-1} V^{abc} \tag{7.5}$$

where A^{-1} is obtained as follows

$$A = \begin{vmatrix} 1 & 1 & 1 \\ 1 & a^2 & a \\ 1 & a & a^2 \end{vmatrix} = 1(a - a^2) - 1(a^2 - a) + 1(a - a^2)$$

$$= a - a^2 - a^2 + a + a - a^2 = 3(a - a^2) \quad \text{since } a^4 = a$$

$$\text{Cofactor matrix} = \begin{pmatrix} a - a^2 & a - a^2 & a - a^2 \\ a - a^2 & a^2 - 1 & 1 - a \\ a - a^2 & 1 - a & a^2 - 1 \end{pmatrix}$$

It is symmetrical about main diagonal. $\therefore$ Transpose is same.

$$A^{-1} = \frac{1}{3(a - a^2)} \begin{pmatrix} a - a^2 & a - a^2 & a - a^2 \\ a - a^2 & a^2 - 1 & 1 - a \\ a - a^2 & 1 - a & a^2 - 1 \end{pmatrix}$$

Since

$$a^2 - 1 = a^2 - a^3 = a(a - a^2)$$
$$1 - a = a^3(1 - a) = a^3 - a^4 \quad \text{since } a^3 = 1$$
$$= a^2(a - a^2)$$

$$A^{-1} = \frac{1}{3} \begin{pmatrix} 1 & 1 & 1 \\ 1 & a & a^2 \\ 1 & a^2 & a \end{pmatrix} \tag{7.6}$$

Writing the matrix equation

$$\begin{pmatrix} V_{a0} \\ V_{a1} \\ V_{a2} \end{pmatrix} = \frac{1}{3} \begin{pmatrix} 1 & 1 & 1 \\ 1 & a & a^2 \\ 1 & a^2 & a \end{pmatrix} \begin{pmatrix} V_a \\ V_b \\ V_c \end{pmatrix} \tag{7.7}$$

From which we can write

$$V_{a0} = \frac{1}{3}(V_a + V_b + V_c)$$

$$V_{a1} = \frac{1}{3}(V_a + aV_b + a^2 V_c)$$

$$V_{a2} = \frac{1}{3}(V_a + a^2 V_b + aV_c)$$

Similarly for unbalanced current

$$I_{a0} = \frac{1}{3}(I_a + I_b + I_c)$$

$$I_{a1} = \frac{1}{3}(I_a + aI_b + a^2 I_c)$$

$$I_{a2} = \frac{1}{3}(I_a + a^2 I_b + aI_c)$$

In a star connected four-wire system, neutral current is

$$I_n = I_a + I_b + I_c = 3I_{a0} \tag{7.8}$$

or

$$I_{a0} = \frac{1}{3}I_n$$

Flow of neutral currents indicates unbalanced condition and zero-sequence current exists in an unbalanced system.

In a three-wire star or delta connected system

$$I_n = 0$$

$$\therefore \qquad\qquad I_{a0} = 0$$

EXAMPLE 7.1 One conductor of a three-phase line is open. The current flowing to the Δ connected load through line a is 10 A. With the current in line a as reference and assuming that line c is open, find the symmetrical components of the line currents.

Solution Figure 7.2 shows the circuit.

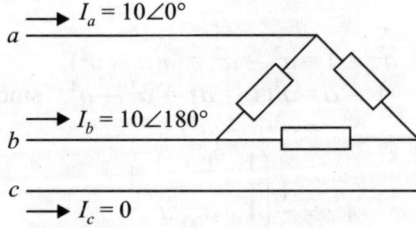

Figure 7.2 Circuit for Example 7.1.

The line currents are

$$I_c = 0; \quad I_a = 10\angle0° \text{ A} \qquad \therefore I_b = -I_a = 10\angle180° \text{ A}$$

$$I_{a0} = \frac{1}{3}(I_a + I_b + I_c) = 0$$

$$I_{a1} = \frac{1}{3}(I_a + aI_b + a^2I_c) = \frac{1}{3}[10\angle0° + 10\angle(180° + 120°) + 0]$$

$$= \frac{1}{3}(10 + 5 - j8.67) = 5 - j2.89 = 5.78\angle-30° \text{ A}$$

$$I_{a2} = \frac{1}{3}(I_a + a^2I_b + aI_c) = \frac{1}{3}[10\angle0° + 10\angle(180° + 240°) + 0]$$

$$= \frac{1}{3}(10 + 5 + j8.67) = 5 + j2.89 = 5.78\angle30° \text{ A}$$

$$I_{b1} = a^2I_{a1} = 5.78\angle210° = 5.78\angle-150° \text{ A}$$
$$I_{b2} = aI_{a2} = 5.78\angle150° \text{ A}$$
$$I_{b0} = I_{a0} = 0 \text{ A}$$
$$I_{c1} = aI_{a1} = 5.78\angle90° \text{ A}$$
$$I_{c2} = a^2I_{a2} = 5.78\angle270° = 5.78\angle-90° \text{ A}$$
$$I_{c0} = I_{a0} = 0 \text{ A}$$

It may be noted that components I_{c1} and I_{c2} have definite values although line c is open and carries no net current.

7.1.5 Power in Terms of Symmetrical Components

Complex power in a three-phase balanced system can be computed as

$$S = 3VI^* \tag{7.9}$$

where V is the phase voltage and I is the phase current.

When the system is unbalanced,

$$S = V_aI_a^* + V_bI_b^* + V_cI_c^* \tag{7.10}$$

In matrix from

$$S = [V_a \quad V_b \quad V_c]\begin{bmatrix} I_a \\ I_b \\ I_c \end{bmatrix}^*$$

$$= [V^{abc}]^T [I^{abc}]^* = [AV^{012}]^T [AI^{012}]^*$$

$$= [V^{012}]^T A^T A^* [I^{012}]^*$$

$$A^T A^* = \begin{pmatrix} 1 & 1 & 1 \\ 1 & a^2 & a \\ 1 & a & a^2 \end{pmatrix}\begin{pmatrix} 1 & 1 & 1 \\ 1 & a & a^2 \\ 1 & a^2 & a \end{pmatrix} \quad \text{since } a^2 = a^* \text{ and vice-versa}$$

$$= 3 \begin{pmatrix} 1 & 0 & 0 \\ 0 & 1 & 0 \\ 0 & 0 & 1 \end{pmatrix} \quad \text{since } 1 + a + a^2 = 0 \text{ and } a^3 = 1 \text{ and } a^4 = a$$

$$= 3U$$

$$\therefore \qquad S = 3[V^{012}]^T [I^{012}]^*$$

$$= 3[V_{a0} \quad V_{a1} \quad V_{a2}] \begin{pmatrix} I_{a0} \\ I_{a1} \\ I_{a2} \end{pmatrix}^*$$

$$S = V_a I_a^* + V_b I_b^* + V_c I_c^* = 3V_{a0} I_{a0}^* + 3V_{a1} I_{a1}^* + 3V_{a2} I_{a2}^* \tag{7.11}$$

EXAMPLE 7.2 A set of three-phase unbalanced voltages and currents resolved into symmetrical components gives the following results:

$$V_{a0} = 30\angle{-30^\circ} \qquad\qquad I_{a0} = 10\angle 190^\circ$$
$$V_{a1} = 450\angle 0^\circ \qquad\qquad I_{a1} = 6\angle 20^\circ$$
$$V_{a2} = 225\angle 40^\circ \qquad\qquad I_{a2} = 5\angle 50^\circ$$

Determine the complex power

Solution

$$S = 3V_{a0} I_{a0}^* + 3V_{a1} I_{a1}^* + 3V_{a2} I_{a2}^*$$
$$= 3 \times 30\angle{-30^\circ} \times 10\angle{-190^\circ} + 3 \times 450\angle 0^\circ \times 6\angle{-20^\circ} + 3 \times 225\angle 40^\circ \times 5\angle{-50^\circ}$$
$$= 900\angle{-220^\circ} + 8100\angle{-20^\circ} + 3375\angle{-10^\circ}$$
$$= 900\angle 140^\circ + 8100\angle{-20^\circ} + 3375\angle{-10^\circ}$$
$$= -689.4 + j578.5 + 7611.5 - j2770.4 + 3323.7 - j586.1$$
$$= 10245.8 - j2778.0$$
$$= 10615.7\angle{-15.2^\circ}$$

7.2 Sequence Impedance

Figure 7.3 shows a transmission system where self impedance of each phase is equal (owing to transposition or proper configuration of lines) and given by Z_s and mutual impedance between any two phases is equal and given by Z_m. Let V_a, V_b and V_c be voltage drops in respective phases.

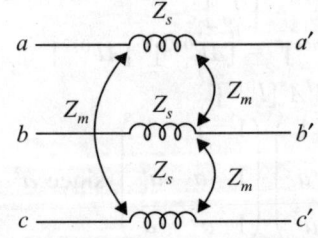

Figure 7.3 Transmission system.

$$\begin{pmatrix} V_a \\ V_b \\ V_c \end{pmatrix} = \begin{pmatrix} Z_s & Z_m & Z_m \\ Z_m & Z_s & Z_m \\ Z_m & Z_m & Z_s \end{pmatrix} \begin{pmatrix} I_a \\ I_b \\ I_c \end{pmatrix}$$

$$V^{abc} = Z^{abc} \, I^{abc}$$

$$A V^{012} = Z^{abc} \, A \, I^{012}$$

$$V^{012} = A^{-1} \, Z^{abc} \, A \, I^{012}$$

$$V^{012} = Z^{012} \, I^{012}$$

where

$$Z^{012} = A^{-1} \, Z^{abc} \, A$$

$$= \frac{1}{3} \begin{pmatrix} 1 & 1 & 1 \\ 1 & a & a^2 \\ 1 & a^2 & a \end{pmatrix} \begin{pmatrix} Z_s & Z_m & Z_m \\ Z_m & Z_s & Z_m \\ Z_m & Z_m & Z_s \end{pmatrix} \begin{pmatrix} 1 & 1 & 1 \\ 1 & a^2 & a \\ 1 & a & a^2 \end{pmatrix}$$

$$= \frac{1}{3} \begin{pmatrix} 1 & 1 & 1 \\ 1 & a & a^2 \\ 1 & a^2 & a \end{pmatrix} \begin{pmatrix} Z_s + 2Z_m & Z_s + (a^2 + a)Z_m & Z_s + (a + a^2)Z_m \\ Z_s + 2Z_m & a^2 Z_s + (1+a)Z_m & aZ_s + (1+a^2)Z_m \\ Z_s + 2Z_m & aZ_s + (a^2 + 1)Z_m & a^2 Z_s + (1+a)Z_m \end{pmatrix}$$

$$= \frac{1}{3} \begin{pmatrix} 1 & 1 & 1 \\ 1 & a & a^2 \\ 1 & a^2 & a \end{pmatrix} \begin{pmatrix} Z_s + 2Z_m & Z_s - Z_m & Z_s - Z_m \\ Z_s + 2Z_m & a^2(Z_s - Z_m) & a(Z_s - Z_m) \\ Z_s + 2Z_m & a(Z_s - Z_m) & a^2(Z_s - Z_m) \end{pmatrix}$$

Since $a^2 + a = -1$, $1 + a = -a^2$, $1 + a^2 = -a$

$$Z^{012} = \frac{1}{3} \begin{pmatrix} 3Z_s + 6Z_m & 0 & 0 \\ 0 & 3Z_s - 3Z_m & 0 \\ 0 & 0 & 3Z_s - 3Z_m \end{pmatrix}$$

$$Z^{012} = \begin{pmatrix} Z_s + 2Z_m & 0 & 0 \\ 0 & Z_s - Z_m & 0 \\ 0 & 0 & Z_s - Z_m \end{pmatrix} \qquad (7.12)$$

From Eq (7.12) for Z^{012}, it may be seen that

1. Symmetrical components of impedance are independent of one another i.e., no coupling exists among zero, positive- and negative-sequence impedances.
2. Positive-and negative-sequence impedances are equal.
3. Zero-sequence impedance is approximately 2.5 times that of positive or negative-sequence impedance, due to different magnetic field set-up by zero-sequence current.
4. If mutual coupling is negligible, all the sequence impedances are equal.

$$V^{012} = Z^{012} \, I^{012} \qquad (7.13)$$

The symmetrical components of unbalanced current flowing in a balanced series impedance or balanced Y connected load produces voltage drops of like sequence only. In other words in any part of the circuit, voltage drop caused by current of a certain sequence depends on the impedance of that part of the circuit to current of that sequence only. The impedance of any section of a balanced network to current of one sequence may be different from impedance to the current of another sequence.

The impedance of the circuit when positive-sequence currents alone are flowing is called the positive-sequence impedance. Similarly, when only negative-sequence currents are present, the impedance is called negative-sequence impedance. When only zero-sequence currents are present, the impedance is called zero-sequence impedance.

7.3   Sequence Network

The analysis of an unsymmetrical fault on a symmetrical system is to find the symmetrical components of the unbalanced currents that are flowing. Since the component currents of one phase sequence cause voltage drops of like sequence only and are independent of currents of other sequences, in a balanced system, currents of any one sequence can be considered to flow in an independent network composed of the impedances to current of that sequence only. The single-phase equivalent circuit composed of the impedances to current of any one sequence only is called the *sequence network* for that particular sequence.

The sequence network includes any generated emfs of like sequence. Sequence networks carrying the sequence currents are interconnected to represent various unbalanced fault conditions. Therefore, to calculate the effect of a fault by the method of symmetrical components, it is essential to determine the sequence impedances and to combine them to form the sequence networks.

The sequence networks for the transmission line is shown in Figure 7.4

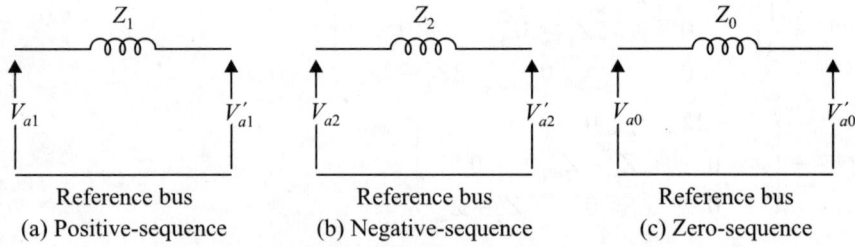

(a) Positive-sequence   (b) Negative-sequence   (c) Zero-sequence

Figure 7.4   Sequence networks for transmission line.

7.3.1   Sequence Network of Unloaded Generator

Circuit diagram of an unloaded generator grounded through a reactance is shown in Figure 7.5.

The generated voltages are of positive sequence only, since the generator is designed to supply balanced three-phase voltages. Therefore, the positive-sequence network is composed of an emf in series with the positive-sequence impedance of the generator, as shown in sequence networks in Figure 7.6(a–f).

$$E_a = E_g = E_g' = E_g''$$

is the positive sequence voltage to neutral (since the generator is unloaded). The reactance in the positive sequence network is the subtransient, transient or synchronous reactance, depending on whether subtransient, transient or steady state condition is being studied.

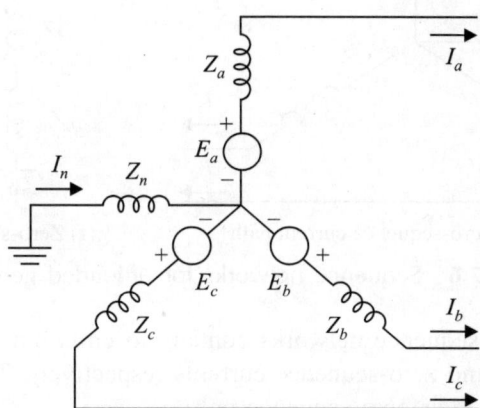

Figure 7.5 Circuit diagram of an unloaded generator.

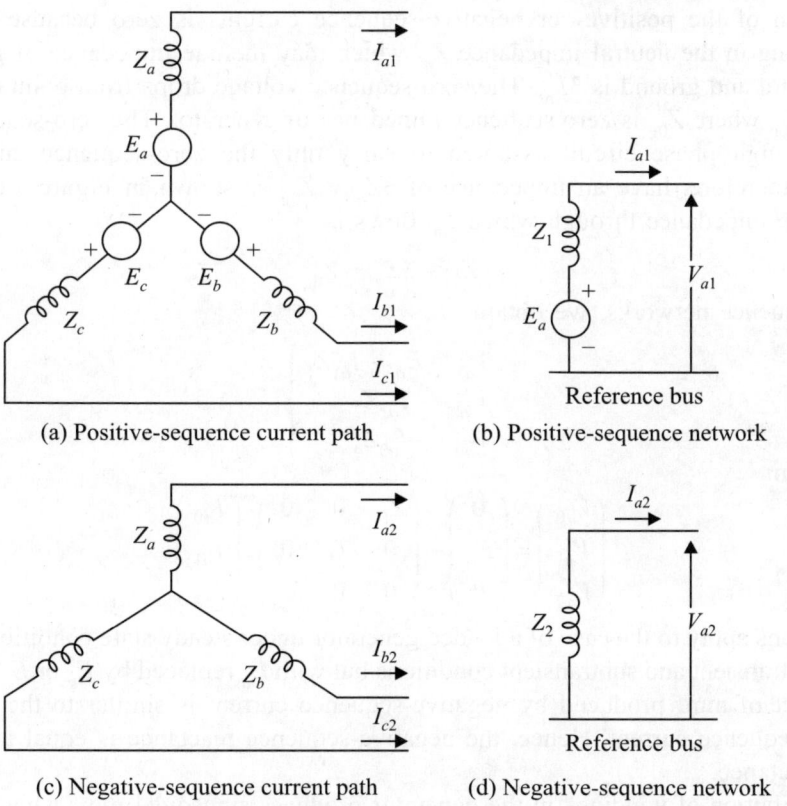

(a) Positive-sequence current path   (b) Positive-sequence network

(c) Negative-sequence current path   (d) Negative-sequence network

Figure 7.6   (*Contd.*)

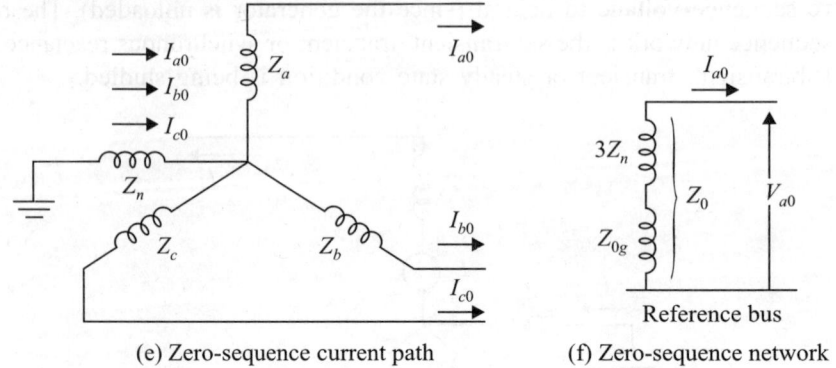

(e) Zero-sequence current path   (f) Zero-sequence network

Figure 7.6 Sequence networks for unloaded generator.

The negative and zero-sequence networks contain no emfs but include the impedances of the generator to negative- and zero-sequence currents respectively. The sequence currents flow through the impedances of their own sequence only.

The reference bus for the positive- and negative-sequence networks is the ground as no positive- or negative-sequence currents flow from ground to the neutral of the generator, since the sum of the positive- or negative-sequence currents is zero because of symmetry. Current flowing in the neutral impedance Z_n, which may include impedance of ground, if any between neutral and ground is $3I_{a0}$. The zero-sequence voltage drops from point a to ground is $-3I_{a0}Z_n - I_{a0}Z_{0g}$ where Z_{0g} is zero sequence impedance of generator. The zero-sequence network which is a single-phase circuit assumed to carry only the zero-sequence current of one-phase must, therefore, have an impedance of $3Z_n + Z_{0g}$ as shown in Figure 7.6(f). The total zero-sequence impedance through which I_{a0} flows is

$$Z_0 = 3Z_n + Z_{0g} \tag{7.14}$$

From the sequence networks, we obtain

$$\left. \begin{aligned} V_{a1} &= E_a - I_{a1}Z_1 \\ V_{a2} &= -I_{a2}Z_2 \\ V_{a0} &= -I_{a0}Z_0 \end{aligned} \right\} \tag{7.15}$$

In matrix from

$$\begin{pmatrix} V_{a0} \\ V_{a1} \\ V_{a2} \end{pmatrix} = \begin{pmatrix} 0 \\ E_a \\ 0 \end{pmatrix} - \begin{pmatrix} Z_0 & 0 & 0 \\ 0 & Z_1 & 0 \\ 0 & 0 & Z_2 \end{pmatrix} \begin{pmatrix} I_{a0} \\ I_{a1} \\ I_{a2} \end{pmatrix} \tag{7.16}$$

These equations apply to the case of a loaded generator under steady state conditions as well and also apply to transient and subtransient conditions but with E_a replaced by E'_g or E''_g, respectively.

The effect of mmf produced by negative-sequence current is similar to the one produced by positive-sequence current. Hence, the negative-sequence reactance is equal to the positive-sequence reactance.

The distribution of windings in the generator produces sinusoidal mmf whose sum is zero. The zero-sequence reactance is therefore due to leakage only and is about 60% of positive-sequence reactance.

7.3.2 Sequence Network of Transformers

The sequence impedances of the transformer are equal. However, the zero sequence network depends on the configuration of primary and secondary, and the connection of neutral to ground.

We know that no current flows in the primary of a transformer unless current flows in the secondary, neglecting the relatively small magnetizing current. Five possible connections of two winding transformers are shown in Figure 7.7. The arrows indicate the flow of zero-sequence current.

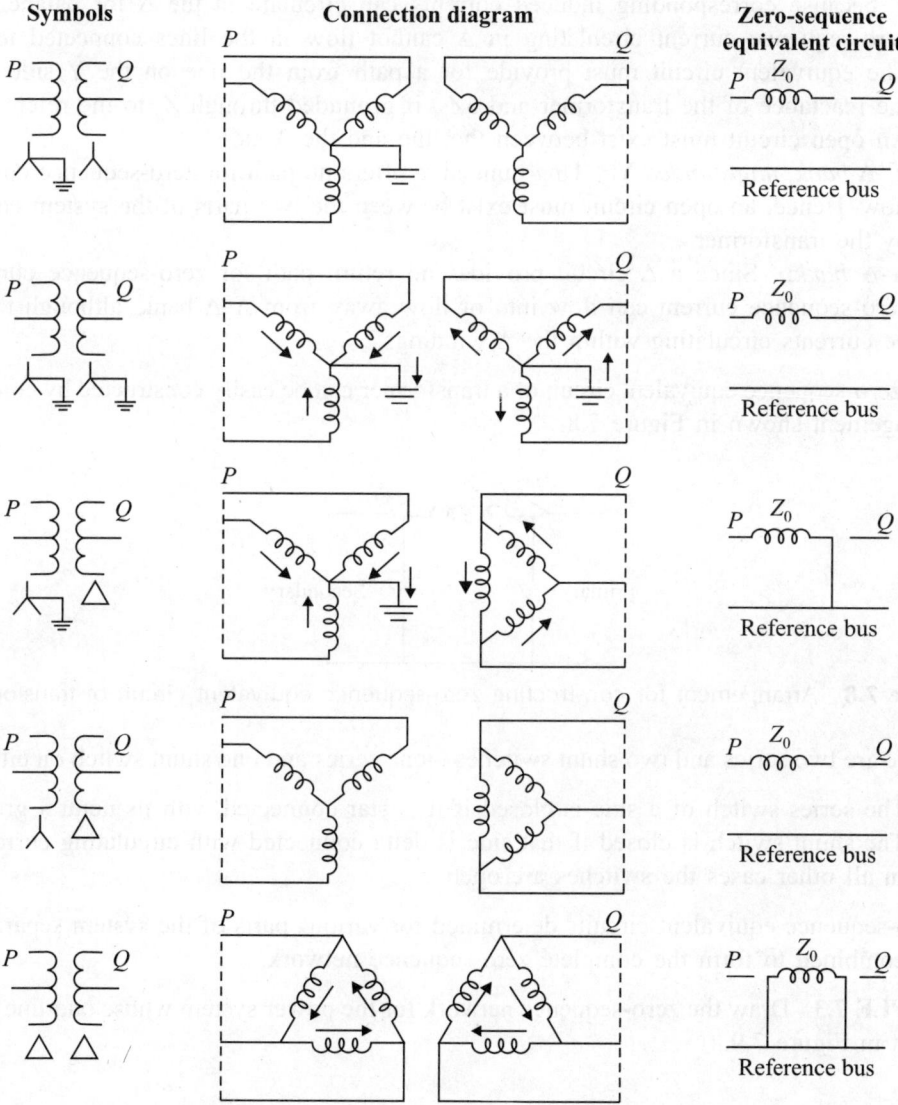

Figure 7.7 Zero-sequence equivalent circuits.

The explanation for the equivalent circuits are as follows:

1. Y–Y *bank, one neutral grounded*: If either of the neutral terminals is ungrounded, there is no path for the flow of zero sequence current. Hence, an open circuit must exist between the two parts of the system connected by the transformer.
2. Y–Y *bank, both neutrals grounded*: There is a path for the zero sequence currents to flow in both windings of the transformer. Parts on the two sides of the transformer are connected through the zero-sequence impedance.
3. Y–Δ *bank, grounded* Y: Zero-sequence currents have a path to ground through the Y because corresponding induced currents can circulate in the Δ for balancing. The zero-sequence current circulating in Δ cannot flow in the lines connected to the Δ. The equivalent circuit must provide for a path from the line on the Y side through the reactance of the transformer and $3Z_n$, if grounded through Z_n to the reference bus. An open circuit must exist between the line and the Δ side.
4. Y–Δ *bank, ungrounded* Y: Ungrounded Y offers no path for zero-sequence currents to flow. Hence, an open circuit must exist between the two parts of the system connected by the transformer.
5. Δ–Δ *bank*: Since a Δ circuit provides no return path for zero-sequence current, no zero-sequence current can flow into or flow away from Δ–Δ bank, although there can be currents circulating within the Δ windings.

The zero-sequence equivalent circuit of a transformer can be easily constructed by considering the arrangement shown in Figure 7.8.

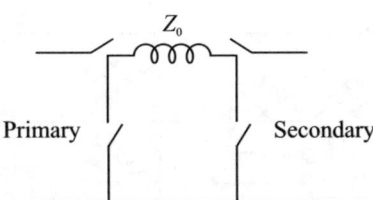

Figure 7.8 Arrangement for constructing zero-sequence equivalent circuit of transformer.

There are two series and two shunt switches—one series and one shunt switch on either side.

- The series switch of a side is closed if it is star connected with its neutral grounded.
- The shunt switch is closed if that side is delta connected with circulating current flow.
- In all other cases the switches are open.

Zero-sequence equivalent circuits determined for various parts of the system separately are readily combined to form the complete zero-sequence network.

EXAMPLE 7.3 Draw the zero-sequence network for the power system whose one-line diagram is shown in Figure 7.9.

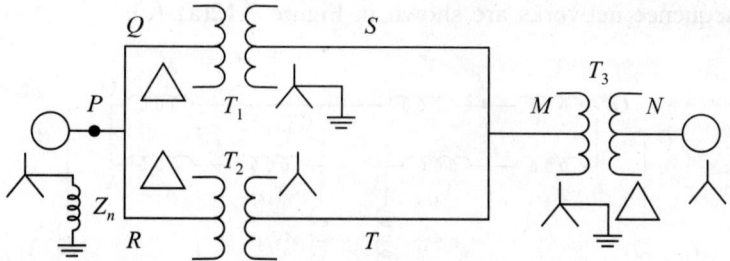

Figure 7.9 One-line diagram for Example 7.3.

Solution The zero-sequence network is shown in Figure 7.10.

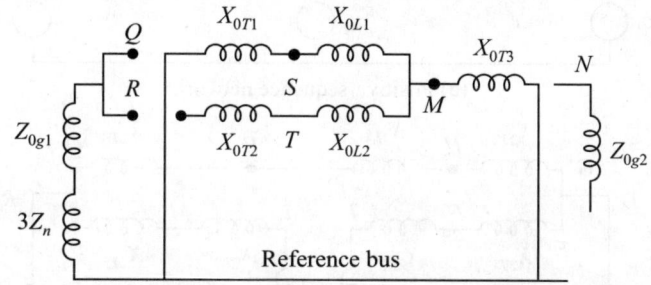

Figure 7.10 Zero-sequence network for Example 7.3.

EXAMPLE 7.4 For the power system shown in Figure 7.11, draw the positive-, negative- and zero-sequence networks.

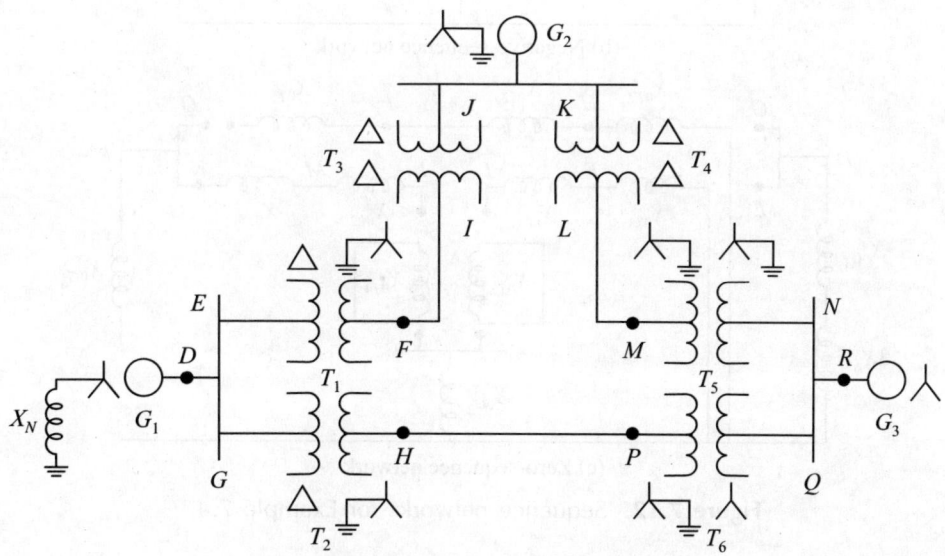

Figure 7.11 System for Example 7.4.

Solution The sequence networks are shown in Figure 7.12(a)–(c).

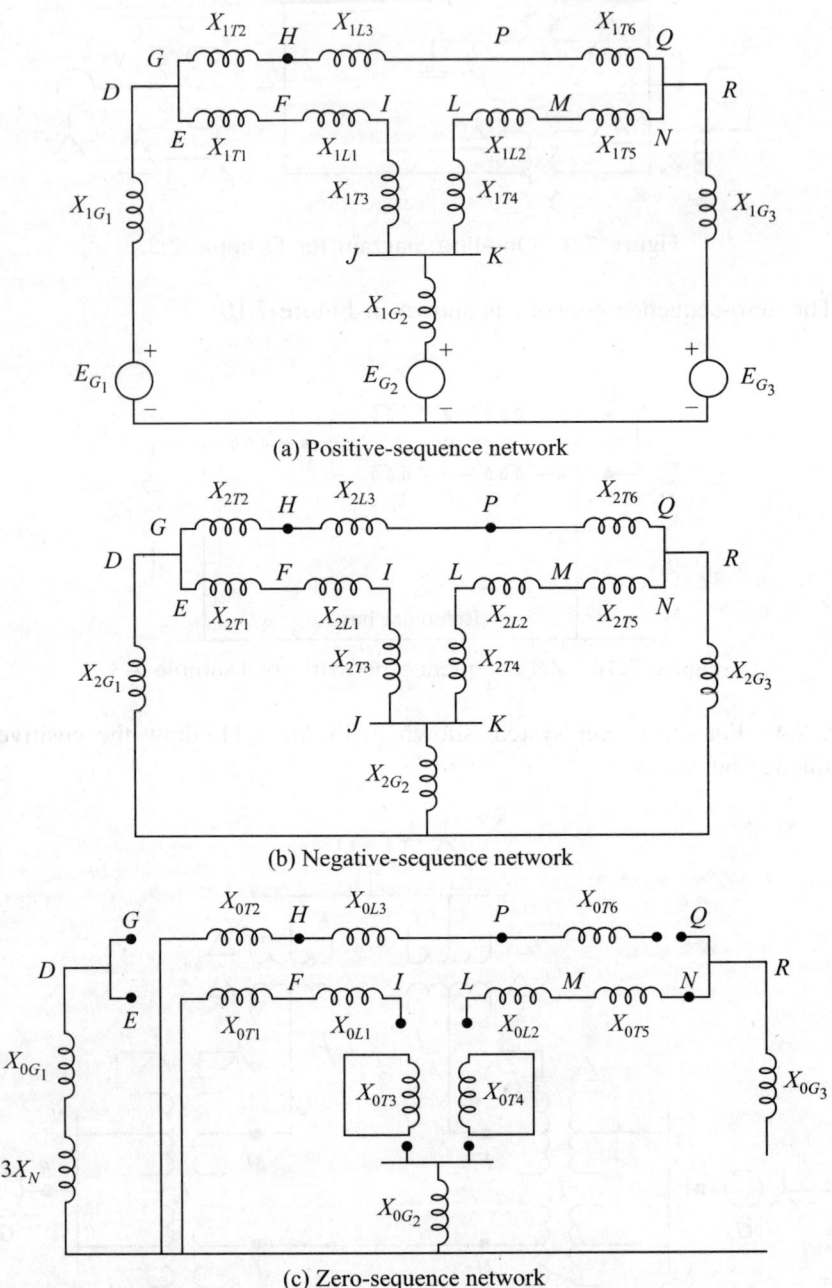

(a) Positive-sequence network

(b) Negative-sequence network

(c) Zero-sequence network

Figure 7.12 Sequence networks for Example 7.4.

EXAMPLE 7.5 Determine the positive-, negative- and zero-sequence networks for the system shown in Figure 7.13. Assume zero-sequence reactances for the generator and synchronous motors as 0.06 pu. Current limiting reactors of 2.5 Ω are connected in the neutral of the generator and motor 2. The zero-sequence reactance of the transmission line is 300 Ω. Find the zero-sequence bus impedance matrix by means of Z_{bus} building algorithm.

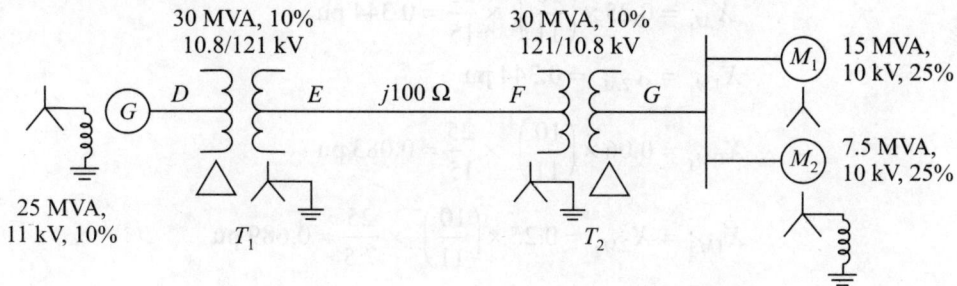

Figure 7.13 One-line diagram for Example 7.5.

Solution Let us select generator ratings as new base values for entire system.

$$S_b = 25 \text{ MVA} \qquad V_b = 11 \text{ kV}$$

Sequence reactances of the generator:

Since generator rating and new base values are same, the generator pu reactances do not change.

$$X_{1G} = 0.1; \qquad X_{2G} = 0.1; \qquad X_{0G} = 0.06$$

$$Z_{bG} = \frac{11^2}{25} = 4.84 \ \Omega$$

$$X_{NG} = \frac{2.5}{4.84} = 0.517 \text{ pu}$$

Sequence reactances of transformer T_1:

$$X_{T_1} = 0.1 \times \left(\frac{10.8}{11}\right)^2 \times \frac{25}{30} = 0.08 \text{ pu}$$

$$X_{1T_1} = X_{2T_1} = X_{0T_1} = 0.08 \text{ pu}$$

Sequence reactances of transmission line

$$V_{bL} = 11 \times \frac{121}{10.8} = 123.24 \text{ kV}$$

$$Z_{bL} = \frac{123.24^2}{25} = 607.5 \ \Omega$$

$$X_L = \frac{100}{607.5} = 0.165 \text{ pu}$$

$$X_{1L} = X_{2L} = 0.165 \text{ pu}$$

$$X_{0L} = \frac{300}{607.5} = 0.495 \text{ pu}$$

Sequence reactances of motors:

$$V_{bM} = V_{bL} \times \frac{10.8}{121} = 11 \times \frac{121}{10.8} \times \frac{10.8}{121} = 11 \, \text{kV}$$

$$X_{M_1} = 0.25 \times \left(\frac{10}{11}\right)^2 \times \frac{25}{15} = 0.344 \, \text{pu}$$

$$X_{1M_1} = X_{2M_1} = 0.344 \, \text{pu}$$

$$X_{0M_1} = 0.06 \times \left(\frac{10}{11}\right)^2 \times \frac{25}{15} = 0.083 \, \text{pu}$$

$$X_{1M_2} = X_{2M_2} = 0.25 \times \left(\frac{10}{11}\right)^2 \times \frac{25}{7.5} = 0.689 \, \text{pu}$$

$$X_{0M_2} = 0.06 \times \left(\frac{10}{11}\right)^2 \times \frac{25}{7.5} = 0.165 \, \text{pu}$$

$$Z_{bM} = \frac{11^2}{25} = 4.84 \, \Omega$$

$$X_{NM} = \frac{2.5}{4.84} = 0.517 \, \text{pu}$$

Sequence reactances of transformer T_1:

$$X_{T_2} = 0.1 \times \left(\frac{10.8}{11}\right)^2 \times \frac{25}{30} = 0.08 \, \text{pu}$$

$$X_{1T_2} = X_{2T_2} = X_{0T_2} = 0.08 \, \text{pu}$$

The sequence networks are shown in Figure 7.14(a)–(c).

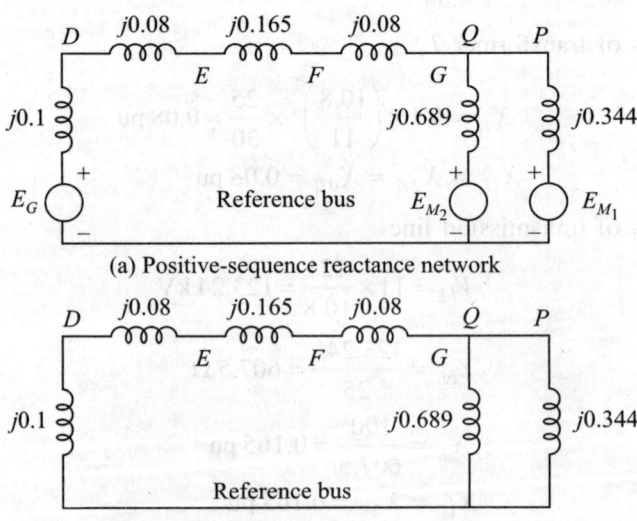

(a) Positive-sequence reactance network

(b) Negative-sequence reactance network

Figure 7.14   (*Contd.*)

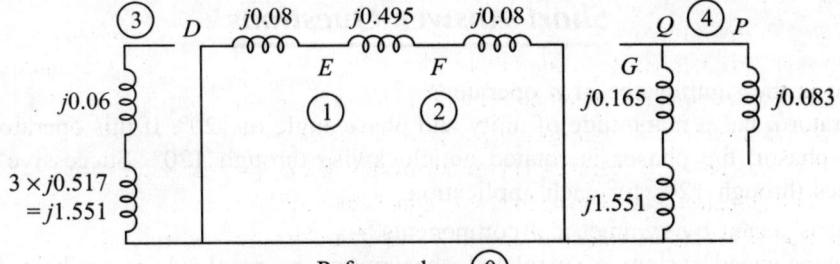

(c) Zero-sequence reactance network

Figure 7.14 Sequence networks for Example 7.5.

Z_{bus} for zero-sequence network

1. Add $\quad z_{10} = j0.08$ from bus (1) (new) to bus (0) (ref)

$$Z_{bus}^0 = 1 \quad (j0.8)$$

2. Add $\quad z_{12} = j0.495$ from bus (2) (new) to bus (1) (exist)

$$Z_{bus}^0 = \begin{array}{c} 1 \\ 2 \end{array}\begin{pmatrix} j0.08 & j0.08 \\ j0.08 & j0.08 + j0.495 \end{pmatrix} = \begin{pmatrix} j0.08 & j0.08 \\ j0.08 & j0.575 \end{pmatrix}$$

3. Add $z_{20} = j0.08$ from bus (2) (exist) to bus (0) reference.

$$Z_{bus}^0 = \begin{pmatrix} j0.08 & j0.08 \\ j0.08 & j0.575 \end{pmatrix} - \frac{1}{j0.575 + j0.08}\begin{pmatrix} j0.08 \\ j0.575 \end{pmatrix}(j0.08 \quad j0.575)$$

$$= \begin{pmatrix} j0.08 & j0.08 \\ j0.08 & j0.575 \end{pmatrix} - \frac{1}{j0.655}\begin{pmatrix} j0.0064 & j0.0460 \\ j0.0460 & j0.3306 \end{pmatrix}$$

$$= \begin{pmatrix} j0.08 & j0.08 \\ j0.08 & j0.575 \end{pmatrix} - \begin{pmatrix} j0.0098 & j0.0702 \\ j0.0702 & j0.5047 \end{pmatrix}$$

$$= \begin{array}{c} 1 \\ 2 \end{array}\begin{pmatrix} j0.0702 & j0.0098 \\ j0.0098 & j0.0703 \end{pmatrix}$$

Buses (3) and (4) are new buses connected directly to reference bus.

$$Z_{bus}^0 = \begin{array}{c} 1 \\ 2 \\ 3 \\ 4 \end{array}\begin{array}{cccc} 1 & 2 & 3 & 4 \\ \begin{pmatrix} j0.0702 & j0.0098 & 0 & 0 \\ j0.0098 & j0.0703 & 0 & 0 \\ 0 & 0 & j1.611 & 0 \\ 0 & 0 & 0 & j1.716 \end{pmatrix} \end{array}$$

Short Answer Questions

1. What is the significance of *a* operator?

Operator *a* has a magnitude of unity and phase angle of 120°. If this operator is applied to a phasor, the phasor is rotated anticlockwise through 120°. Successive application rotates through 120° for each application.

2. What is meant by symmetrical components?

An unbalanced system of *n* related phasors can be resolved into *n* balanced phasors called symmetrical components of the original unbalanced phasors. The symmetrical components are equal in magnitude and have equal phase angle between adjacent phasors.

3. Write the symmetrical components of three-phase system.

In a three-phase system, the three unbalanced phasors (either voltage or current) can be resolved into three balanced system of phasors called positive-sequence components, negative-sequence components and zero-sequence components.

4. What are positive-sequence components of three-phase system?

The positive-sequence components of a three-phase unbalanced phasors consist of three phasors of equal magnitude, displaced from each other by 120° in phase and having same phase sequence as the original phasors.

5. What are negative-sequence components of three-phase system?

The negative-sequence components of a three-phase unbalanced phasors consist of three phasors of equal magnitude, displaced from each other by 120° in phase and having phase sequence opposite to that of the original phasors.

6. What are zero-sequence components of three-phase system?

The zero-sequence components of a three-phase unbalanced phasors consist of three phasors equal in magnitude with zero phase displacement from each other.

7. Express the value of *a* in both polar and rectangular coordinate forms.

$$a = 1\angle 120° = 1\ e^{j2\pi/3}\text{---polar form}$$
$$= -0.5 + j0.866\text{---rectangular form}$$

8. Express the unbalanced voltages in terms of symmetrical components of phase *a*.

$$\begin{pmatrix} V_a \\ V_b \\ V_c \end{pmatrix} = \begin{pmatrix} 1 & 1 & 1 \\ 1 & a^2 & a \\ 1 & a & a^2 \end{pmatrix} \begin{pmatrix} V_{a0} \\ V_{a1} \\ V_{a2} \end{pmatrix}$$

9. Express the symmetrical components of phase *a* in terms of unbalanced voltages.

$$\begin{pmatrix} V_{a0} \\ V_{a1} \\ V_{a2} \end{pmatrix} = \frac{1}{3} \begin{pmatrix} 1 & 1 & 1 \\ 1 & a & a^2 \\ 1 & a^2 & a \end{pmatrix} \begin{pmatrix} V_a \\ V_b \\ V_c \end{pmatrix}$$

10. What are sequence impedance and sequence network?

The sequence impedance is the impedance offered by the device for the like sequence component of the current.

The single-phase equivalent circuit of a power system consists of impedances to current of any one sequence only is called sequence network.

11. What are positive-, negative- and zero-sequence impedances?

The impedance of a circuit element for positive-, negative- and zero-sequence component currents are called positive-, negative- and zero-sequence impedances, respectively.

12. What are positive-, negative- and zero-sequence reactance diagrams?

The reactance diagrams of a power system, when formed using positive-, negative- and zero-sequence reactances are called positive-, negative- and zero-sequence reactance diagrams, respectively.

Exercises

7.1 Two synchronous machines are connected through three-phase transformers to the transmission line as shown Figure E7.1.

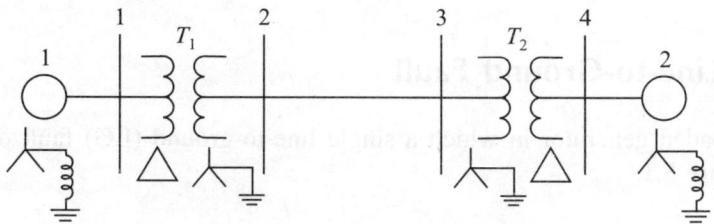

Figure E7.1   Circuit for Problem 7.1.

The ratings and reactances of the machines and transformers are

Machine 1 and 2:   100 MVA, 20 kV, $X_d'' = X_1 = X_2 = 20\%$, $X_0 = 4\%$, $X_n = 5\%$.

Transformers T_1 and T_2:   100 MVA, 20 Δ/345 Y kV, $X = 8\%$.

On a chosen base of 100 MVA, 345 kV in the transmission line circuit the line reactances are $X_1 = X_2 = 15\%$ and $X_0 = 50\%$. Draw each of the sequence networks and find the zero-sequence bus impedance matrix by means of Z_{bus} building algorithm.

$$\textbf{Ans.} \begin{bmatrix} j0.19 & 0 & 0 & 0 \\ 0 & j0.08 & j0.08 & 0 \\ 0 & j0.08 & j0.58 & 0 \\ 0 & 0 & 0 & j0.19 \end{bmatrix}$$

7.2 Draw the zero sequence equivalent circuit for the three-phase transformer connections shown below.

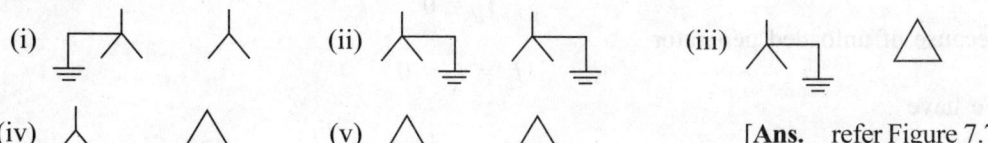

[**Ans.**   refer Figure 7.7]

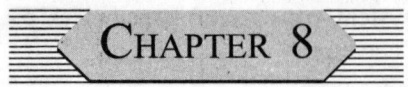

Unbalanced Fault Analysis

Method of symmetrical components is used to resolve unbalanced system into a balanced system. Thus per phase analysis can be extended to unbalanced system also.

8.1 Single Line-to-Ground Fault

Consider an unloaded generator in which a single line-to-ground (LG) fault occurs at phase a as shown in Figure 8.1.

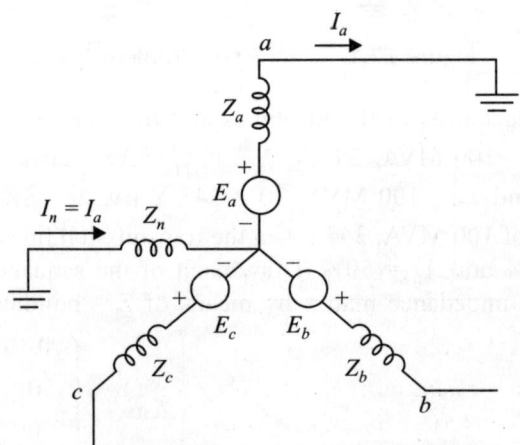

Figure 8.1 Single line-to-ground fault in an unloaded generator.

Due to fault

$$V_a = 0$$

Because of unloaded generator

$$I_b = I_c = 0$$

We have

$$I^{012} = A^{-1} I^{abc}$$

With $I_b = 0$ and $I_c = 0$, the symmetrical components are given by

$$\begin{pmatrix} I_{a0} \\ I_{a1} \\ I_{a2} \end{pmatrix} = \frac{1}{3} \begin{pmatrix} 1 & 1 & 1 \\ 1 & a & a^2 \\ 1 & a^2 & a \end{pmatrix} \begin{pmatrix} I_a \\ 0 \\ 0 \end{pmatrix} = \frac{1}{3} \begin{pmatrix} I_a \\ I_a \\ I_a \end{pmatrix}$$

$$\therefore \qquad\qquad I_{a0} = I_{a1} = I_{a2} = \frac{I_a}{3} \qquad\qquad\qquad (8.1)$$

or fault current,

$$I_f = I_n = I_a = 3I_{a1} \qquad\qquad\qquad (8.2)$$

The connected sequence network for the LG fault satisfying the conditions

$$I_{a0} = I_{a1} = I_{a2} \quad \text{and} \quad V_a = V_{a0} + V_{a1} + V_{a2} = 0$$

is shown in Figure 8.2 for a solid fault.

$$I_{a1} = \frac{E_a}{Z_0 + Z_1 + Z_2} \qquad\qquad\qquad (8.3)$$

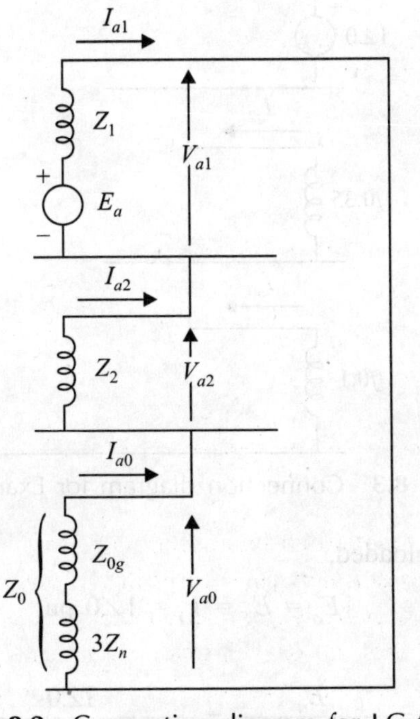

Figure 8.2 Connection diagram for LG fault.

From Eq. (8.2), we get

$$I_f = \frac{3E_a}{Z_0 + Z_1 + Z_2} \qquad\qquad\qquad (8.4)$$

If the neutral is ungrounded, then Z_n will be infinite and hence I_{a1} and the fault current would be zero.

From the knowledge of symmetrical components, the phase and line voltages can be computed.

EXAMPLE 8.1 A salient pole generator without dampers is rated 20 MVA, 13.8 kV and has a direct axis subtransient reactance of 0.25 pu. The negative and zero-sequence reactances are 0.35 and 0.10 pu respectively. The neutral of the generator is solidly grounded. Determine the subtransient current in the generator and the line-to-line voltages for subtransient conditions, when a single line-to-ground fault occurs at the generator terminals with the generator operating unloaded at rated voltage. Neglect resistance.

Solution Connection diagram for the LG fault is shown in Figure 8.3.

$$S_b = 20 \text{ MVA}; \qquad V_{\text{base}} = 13.8 \text{ kV}; \qquad V_{\text{base ph}} = \frac{13.8}{\sqrt{3}} \text{ kV}$$

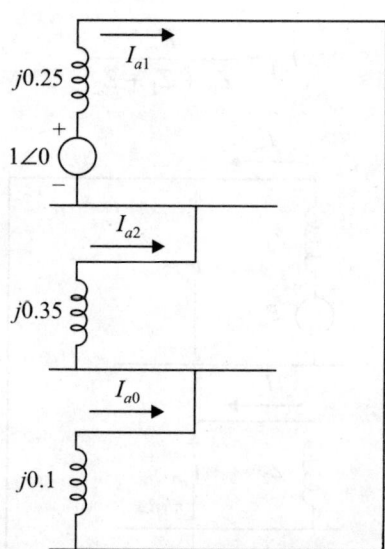

Figure 8.3 Connection diagram for Example 4.6.

Since the generator is unloaded,

$$E_a = E_g = V_t = 1\angle 0 \text{ pu}$$

For an LG fault,

$$I_{a1} = I_{a2} = I_{a0} = \frac{E_a}{Z_0 + Z_1 + Z_2} = \frac{1\angle 0}{j0.1 + j0.35 + j0.25} = -j1.428 \text{ pu}$$

$$I_f = I_a = 3I_{a1} = -j4.29 \text{ pu and } I_b = I_c = 0$$

$$I_{\text{base}} = \frac{20}{\sqrt{3} \times 13.8} = 0.837 \text{ kA}$$

$\therefore$ $$I_f = -j0.429 \times 0.837 = -j3.59 \text{ kA}$$

The symmetrical components of the voltage from point a to ground are

$$V_{a1} = E_a - I_{a1}Z_1 = 1 - (-j1.428) \times j0.25$$
$$= 1 - j0.3575 = 0.6425 \text{ pu}$$
$$V_{a2} = -I_{a2}Z_2 = -(-j1.428) \times j0.35 = -0.5 \text{ pu}$$
$$V_{a0} = -I_{a0}Z_0 = -(-j1.428) \times j0.1 = -0.1425 \text{ pu}$$

Line-to-ground voltages are

$$V_a = 0$$
$$V_b = V_{a0} + a^2 V_{a1} + a V_{a2}$$
$$= -0.1425 + 0.6425\angle 240° - 0.5\angle 120°$$
$$= -0.1425 - 0.3213 - j0.5564 + 0.25 - j0.433$$
$$= -0.2143 - j0.9894 \text{ pu}$$
$$V_c = V_{a0} + a V_{a1} + a^2 V_{a2}$$
$$= -0.1425 + 0.6425\angle 120° - 0.5\angle 240°$$
$$= -0.1425 - 0.3213 + j0.5564 - 0.25 + j0.433$$
$$= -0.2143 + j0.9894 \text{ pu}$$

Line-to-line voltages:

$$V_{ab} = V_a - V_b = 0.2143 + j0.9894 = 1.012\angle 77.78°$$
$$V_{bc} = V_b - V_c = -j1.9788 = 1.9788\angle -90°$$
$$V_{ca} = V_c - V_a = -0.2143 + j0.9894 = 1.012\angle 102.22°$$

Since the generated voltage-to-neutral E_a was taken as 1 pu, the above line-to-line voltages are also expressed in per unit of the base voltage to neutral.

$$V_{ab} = 1.012\angle 77.78° \times \frac{13.8}{\sqrt{3}} = 8.06\angle 77.78° \text{ kV}$$

$$V_{bc} = 1.9788\angle -90° \times \frac{13.8}{\sqrt{3}} = 15.764\angle -90° \text{ kV}$$

$$V_{ca} = 1.012\angle 102.22° \times \frac{13.8}{\sqrt{3}} = 8.06\angle 102.22° \text{ kV}$$

EXAMPLE 8.2 An unloaded star connected solidly grounded 10 MVA, 11 kV generator has positive-, negative- and zero-sequence impedances as $j1.3 \ \Omega$, $j1.8 \ \Omega$, and $j0.4 \ \Omega$, respectively. Single line-to-ground fault occurs at the terminals of the generator.

(i) Calculate the fault current.
(ii) Determine the value of inductive reactance that must be inserted at the generator neutral to limit the fault current to 50% of the value obtained in (i).

Solution

$$S_b = 10 \text{ MVA}, \quad V_{base} = 11 \text{ kV}$$
$$Z_b = \frac{(11)^2}{10} = 12.1 \ \Omega$$

$$I_{base} = \frac{10}{\sqrt{3} \times 11} = 0.5249 \text{ kA}$$

Since the generator is unloaded,

$$E_a = E_g = V_t = 1\angle 0° \text{ pu}$$

(i) For LG fault with $Z_n = 0$

$$I_{a1} = I_{a2} = I_{a0} = \frac{E_a}{Z_0 + Z_1 + Z_2} = \frac{1\angle 0°}{(j0.4 + j1.8 + j1.3)/12.1} = -j3.4571 \text{ pu}$$

$$I_f = I_a = 3I_{a1} = -j10.3713 \text{ pu}$$

$$\therefore \qquad I_f = 10.3713 \times 0.5249 = 5.4439 \text{ kA}$$

(ii)
$$Z_0 = Z_{0g} + 3Z_n$$

$$\frac{3E_a}{Z_0 + Z_1 + Z_2} = 0.5 \times \frac{3E_a}{Z_{0g} + Z_1 + Z_2}$$

$$(Z_0 + Z_1 + Z_2) = 2 (Z_{0g} + Z_1 + Z_2)$$

$$Z_{0g} + 3Z_n + Z_1 + Z_2 = 2 (Z_{0g} + Z_1 + Z_2)$$

$$3Z_n = Z_{0g} + Z_1 + Z_2 = j0.4 + j1.3 + j1.8$$

$$Z_n = \frac{j3.5}{3} = j1.167 \text{ } \Omega$$

8.2 Line-to-Line Fault

Consider a line-to-line (LL) fault between phases b and c on an unloaded Y connected generator as shown in Figure 8.4.

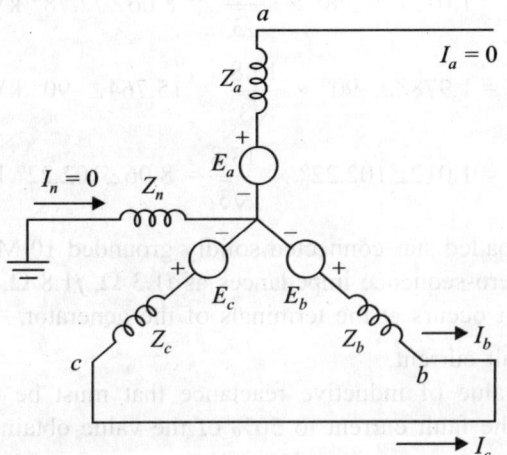

Figure 8.4 Line-to-line fault on an unloaded generator.

Due to fault $I_c = -I_b$, $V_c = V_b$
Because of unloaded generator $I_a = 0$

With $V_c = V_b$, the symmetrical components of voltage are

$$V^{012} = A^{-1}V^{abc}$$

$$\begin{pmatrix} V_{a0} \\ V_{a1} \\ V_{a2} \end{pmatrix} = \frac{1}{3}\begin{pmatrix} 1 & 1 & 1 \\ 1 & a & a^2 \\ 1 & a^2 & a \end{pmatrix}\begin{pmatrix} V_a \\ V_b \\ V_c \end{pmatrix}$$

From which we find

$$V_{a1} = \frac{1}{3}(V_a + aV_b + a^2V_b)$$

$$V_{a2} = \frac{1}{3}(V_a + a^2V_b + aV_b)$$

$$V_{a1} = V_{a2} \tag{8.5}$$

With $I_c = -I_b$ and $I_a = 0$, the symmetrical components of current are

$$I^{012} = A^{-1}I^{abc}$$

$$\begin{pmatrix} I_{a0} \\ I_{a1} \\ I_{a2} \end{pmatrix} = \frac{1}{3}\begin{pmatrix} 1 & 1 & 1 \\ 1 & a & a^2 \\ 1 & a^2 & a \end{pmatrix}\begin{pmatrix} 0 \\ I_b \\ -I_b \end{pmatrix}$$

From this, we have

$$I_{a0} = 0$$

$$I_{a1} = \frac{1}{3}(aI_b - a^2I_b) = \frac{1}{3}(a-a^2)I_b; \tag{8.6}$$

$$I_{a2} = \frac{1}{3}(a^2I_b - aI_b) = \frac{1}{3}(a^2 - a)I_b$$

$$I_{a1} = -I_{a2} \tag{8.7}$$

Since $I_{a0} = 0$ and Z_0 is finite, $V_{a0} = 0$

Since $V_{a1} = V_{a2}$ and $I_{a1} = -I_{a2}$, positive- and negative-sequence networks must be connected in parallel. Since $I_{a0} = 0$, zero-sequence network will not be present. Accordingly, the connection diagram for line-to-line fault is shown in Figure 8.5.

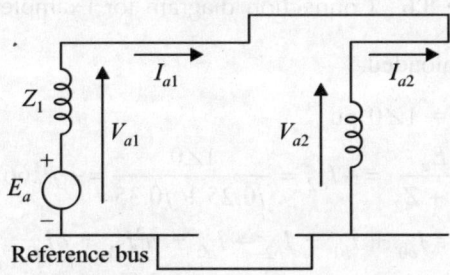

Figure 8.5 Connection diagram for line-to-line fault.

$$E_a - I_{a1}Z_1 = I_{a1}Z_2$$

$$I_{a1} = \frac{E_a}{Z_1 + Z_2} \qquad (8.8)$$

From Eq. (8.6), we obtain

$$I_{a1} = \frac{1}{3}(a - a^2)I_b$$

$$= \frac{1}{3}\left[-0.5 + j\frac{\sqrt{3}}{2} - \left(-0.5 - j\frac{\sqrt{3}}{2}\right)\right]I_b$$

$$= \left(\frac{-j}{\sqrt{3}}\right)I_b$$

or

$$I_f = I_b = -j\sqrt{3}I_{a1} \qquad (8.9)$$

$$I_f = -j\sqrt{3}\frac{E_a}{Z_1 + Z_2} \qquad \text{[using Eq. (8.8)]} \qquad (8.10)$$

EXAMPLE 8.3 Find the subtransient currents and the line-to-line voltages at the fault when a line-to-line fault occurs at the terminals of the generator described in Example 8.1.

Solution Connection diagram for the LL fault is shown in Figure 8.6.

$$S_b = 20 \text{ MVA}, \qquad V_{\text{base}} = 13.8 \text{ kV}, \qquad V_{\text{base ph}} = \frac{13.8}{\sqrt{3}} \text{ kV}$$

$$I_{\text{base}} = \frac{20}{\sqrt{3} \times 13.8} = 0.837 \text{ kA}$$

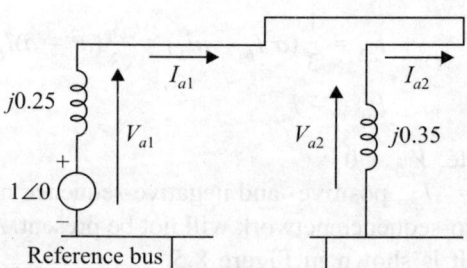

Figure 8.6 Connection diagram for Example 8.3.

Since the generator is unloaded,

$$E_a = E_g = 1\angle 0 \text{ pu}$$

$$I_{a1} = \frac{E_a}{Z_1 + Z_2} = -I_{a2} = \frac{1\angle 0}{j0.25 + j0.35} = -j1.667$$

$$I_f = I_b = I_{b0} + I_{b1} + I_{b2} = I_{a0} + a^2 I_{a1} + a I_{a2}$$

$$= 0 + (a^2 - a)I_{a1} = \left[-0.5 - j\frac{\sqrt{3}}{2} - \left(-0.5 + j\frac{\sqrt{3}}{2}\right)\right]I_{a1}$$

$$= - j\sqrt{3}I_{a1} = - j\sqrt{3}\frac{E_a}{Z_1 + Z_2}$$

$$= - j\sqrt{3}\frac{1\angle 0°}{j0.25 + j0.35} = - j2.886 \text{ pu}$$

and
$$I_c = - I_b = j2.886 \text{ pu}, \quad I_a = 0$$
$$I_f = 2.886 \times 0.837 = 2.416 \text{ kA}$$

The symmetrical components of the voltage from a to ground

$$V_{a0} = 0 \quad \text{since } I_{a0} = 0 \text{ and } Z_0 \text{ is finite}$$
$$V_{a1} = V_{a2} = -I_{a2}Z_2 = I_{a1}Z_2 = -j1.667 \times j0.35$$
$$= 0.5833 \text{ pu}$$

Line-to-ground voltages:

$$V_a = V_{a0} + V_{a1} + V_{a2} = 2 \times 0.5833 = 1.1666 \text{ pu}$$
$$V_b = V_{b0} + a^2 V_{a1} + a V_{a2} = (a^2 + a) V_{a1} = -V_{a1} \text{ since } a^2 + a = -1$$
$$= -0.5833 \text{ pu}$$
$$V_c = V_b = - 0.5833 \text{ pu}$$

Line-to-line voltages:

$$V_{ab} = V_a - V_b = 1.7499 \text{ pu} = 1.7499 \times \frac{13.8}{\sqrt{3}} = 13.9422 \text{ kV}$$

$$V_{bc} = V_b - V_c = 0$$

$$V_{ca} = V_c - V_a = -1.7499 \text{ pu} = -1.7499 \times \frac{13.8}{\sqrt{3}} = -13.9422 \text{ kV}$$

8.3 Double Line-to-Ground Fault

Consider a double line-to-ground (LLG) fault involving phases b and c on an unloaded generator as shown in Figure 8.7.

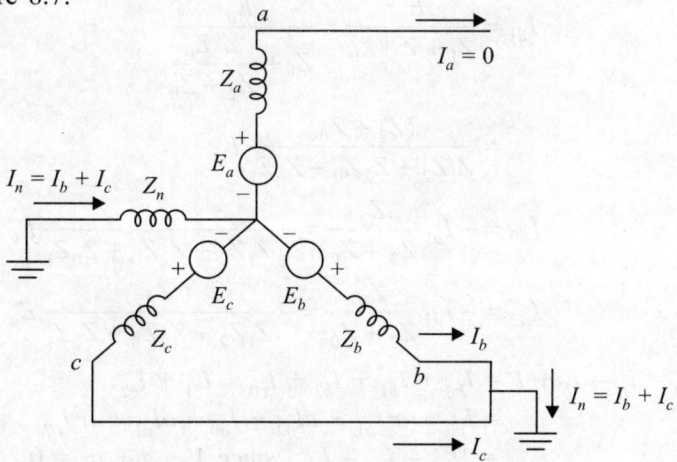

Figure 8.7 Double line-to-ground fault on an unloaded generator.

Due to fault

$$V_b = V_c = 0$$

Because of unloaded generator,

$$I_a = 0$$

i.e.,

$$I_{a0} + I_{a1} + I_{a2} = 0$$

With $V_b = V_c = 0$, the symmetrical components of voltage are

$$V^{012} = A^{-1}V^{abc}$$

$$\begin{pmatrix} V_{a0} \\ V_{a1} \\ V_{a2} \end{pmatrix} = \frac{1}{3} \begin{pmatrix} 1 & 1 & 1 \\ 1 & a & a^2 \\ 1 & a^2 & a \end{pmatrix} \begin{pmatrix} V_a \\ 0 \\ 0 \end{pmatrix}$$

$$V_{a0} = V_{a1} = V_{a2} = \frac{V_a}{3} \tag{8.11}$$

The condition that sum of the sequence currents being zero and sequence voltages being equal is realized by the parallel connection of the sequence networks as shown in Figure 8.8.

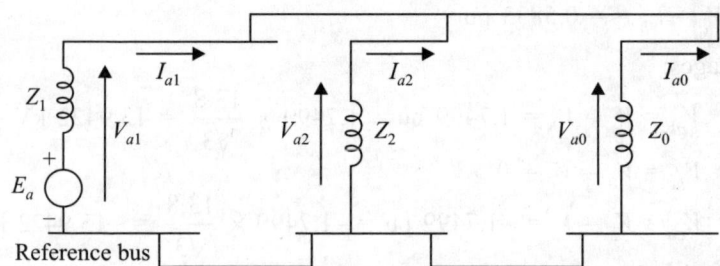

Figure 8.8 Connection diagram for LLG fault.

The sequence currents can be computed from the connection diagram.

$$I_{a1} = \frac{E_a}{Z_1 + Z_2 \| Z_0} = \frac{E_a}{Z_1 + \dfrac{Z_2 Z_0}{Z_2 + Z_0}}$$

$$= \frac{(Z_2 + Z_0)}{Z_1 Z_2 + Z_2 Z_0 + Z_0 Z_1} E_a$$

$$I_{a0} = -I_{a1} \frac{Z_2}{Z_2 + Z_0} = -\frac{Z_2}{Z_1 Z_2 + Z_2 Z_0 + Z_0 Z_1} E_a$$

$$I_{a2} = -I_{a1} \frac{Z_0}{Z_2 + Z_0} = -\frac{Z_0}{Z_1 Z_2 + Z_2 Z_0 + Z_0 Z_1} E_a$$

$$I_f = I_n = I_b + I_c = I_{b0} + I_{b1} + I_{b2} + I_{c0} + I_{c1} + I_{c2}$$
$$= I_{a0} + a^2 I_{a1} + a I_{a2} + I_{a0} + a I_{a1} + a^2 I_{a2}$$
$$= 2I_{a0} - I_{a1} - I_{a2} \quad \text{since } 1 + a + a^2 = 0$$
$$= 3I_{a0} \quad \text{since } I_{a0} + I_{a1} + I_{a2} = 0$$

$$I_f = -\frac{3Z_2}{Z_1 Z_2 + Z_2 Z_0 + Z_0 Z_1} E_a \qquad (8.12)$$

$$V_a = 3V_{a0} = -3I_{a0}Z_0 = -\frac{3Z_2}{Z_1 Z_2 + Z_2 Z_0 + Z_0 Z_1} E_a \qquad (8.13)$$

EXAMPLE 8.4 Find the subtransient currents and the line-to-line voltages at the fault when a double line-to-ground fault occurs at the terminals of the generator described in Example 8.1.

Solution Connection diagram for the LLG fault is shown in Figure 8.9.

$$S_b = 20 \text{ MVA}, \qquad V_{\text{base}} = 13.8 \text{ kV}, \qquad V_{\text{base ph}} = \frac{13.8}{\sqrt{3}} \text{ kV}$$

$$I_{\text{base}} = \frac{20}{\sqrt{3} \times 13.8} = 0.837 \text{ kA}$$

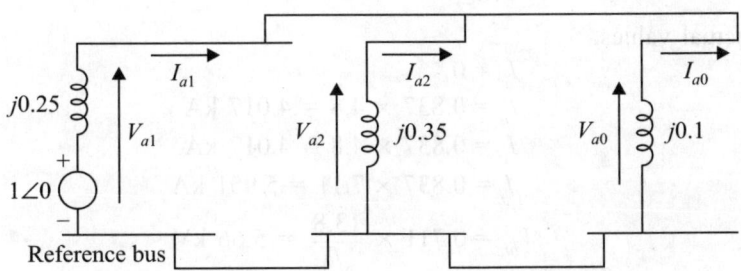

Figure 8.9 Connection diagram for Example 8.4.

Since the generator is unloaded,

$$E_a = E_g = V_t = 1\angle 0° \text{ pu}$$

$$I_{a1} = \frac{E_a}{Z_1 + \dfrac{Z_2 Z_0}{Z_2 + Z_0}} = \frac{1\angle 0°}{j0.25 + \dfrac{j0.35 \times j0.1}{j0.35 + j0.1}}$$

$$= \frac{1}{j0.25 + j0.0778} = \frac{1}{j0.3278}$$

$$= -j3.05 \text{ pu} = 3.05\angle{-90°}$$

$$I_{a0} = -I_{a1} \frac{Z_2}{Z_2 + Z_0} = j3.05 \frac{j0.35}{j0.35 + j0.1}$$

$$= j2.37 \text{ pu} = 2.37\angle 90°$$

$$I_{a2} = -(I_{a1} + I_{a0}) = j0.68 \text{ pu}$$

$$I_a = 0$$

$$I_b = I_{a0} + a^2 I_{a1} + a I_{a2}$$

$$= j2.37 + 3.05\angle 150° + 0.68\angle 210°$$

$$= j2.37 - 2.641 + j1.525 - 0.589 - j0.34$$
$$= -3.23 + j3.555 = 4.8\angle 132.3° \text{ pu}$$
$$I_c = I_{a0} + aI_{a1} + a^2I_{a2}$$
$$= j2.37 + 3.05\angle 30° + 0.68\angle -30°$$
$$= j2.37 + j2.641 + j1.525 + 0.589 - j0.34$$
$$= 3.23 + j3.555 = 4.8\angle 47.7° \text{ pu}$$
$$I_f = I_b + I_c = j7.110 \text{ pu}$$
$$V_{a1} = V_{a2} = -I_{a2}Z_2 = -j0.68 \times j0.35 = 0.237 \text{ pu}$$
$$V_a = V_{a0} + V_{a1} + V_{a2} = 3V_{a2} = 3 \times 0.237 = 0.711 \text{ pu}$$
$$V_b = V_c = 0$$
$$V_{ab} = V_a - V_b = 0.711 \text{ pu}$$
$$V_{bc} = 0$$
$$V_{ca} = V_c - V_a = -0.711 \text{ pu}$$

Expressing in actual values,

$$I_a = 0$$
$$I_b = 0.837 \times 4.8 = 4.017 \text{ kA}$$
$$I_c = 0.837 \times 4.8 = 4.017 \text{ kA}$$
$$I_f = 0.837 \times 7.11 = 5.951 \text{ kA}$$
$$V_{ab} = 0.711 \times \frac{13.8}{\sqrt{3}} = 5.66 \text{ kV}$$
$$V_{bc} = 0$$
$$V_{ca} = 0.711 \times \frac{13.8}{\sqrt{3}} = 5.66 \text{ kV}$$

EXAMPLE 8.5 An alternator of negligible resistance having 1 pu voltage behind transient reactance is subjected to different types of solid fault at its terminals. The pu values of the magnitude of fault current are

   (i) Three-phase fault current = 5 pu
  (ii) LL fault current   = 4 pu
 (iii) LG fault current   = 6 pu

Determine the pu values of the sequence reactances of the machines.

Solution

$$I_{f3ph} = \frac{E_a}{Z_1}$$

$$Z_1 = \frac{E_a}{I_{f3ph}} = \frac{1}{5} = 0.2 \text{ pu}$$

$$I_{f\,LL} = \sqrt{3}\,\frac{E_a}{Z_1 + Z_2}$$

$$Z_1 + Z_2 = \frac{\sqrt{3}E_a}{I_{f\,LL}} = \frac{\sqrt{3} \times 1}{4} = 0.433$$

$$Z_2 = 0.433 - 0.2 = 0.233 \text{ pu}$$

$$I_{f\,LG} = \frac{3E_a}{Z_0 + Z_1 + Z_2}$$

$$Z_1 + Z_2 + Z_0 = \frac{3E_a}{I_{f\,LG}} = \frac{3 \times 1}{6} = 0.5$$

$$Z_0 = 0.5 - 0.433 = 0.067 \text{ pu}$$

8.4   Unbalanced Faults on Power Systems

The analysis of unbalanced faults on power systems is similar to that of such fault on unloaded generator except that each of the sequence networks is replaced by its Thevenin equivalent between the fault point and reference bus and E_a by the pre-fault voltage V_f at the fault point.

EXAMPLE 8.6   A 50 Hz, 50 MVA, 13.2 kV star grounded alternator is connected to a Δ–Y transformer as shown in Figure 8.10. The positive-, negative- and zero-sequence impedances of the alternator are 0.1, 0.1 and 0.05 pu, respectively and that of transformer rated 13.2 kV Δ/120 kV Y 80 MVA with Y solidly grounded 0.1, 0.1 and 0.1 pu. Determine the fault current for a (i) three-phase, (ii) LG fault, (iii) LL fault and (iv) LLG fault at point P. Draw the connection diagram for the sequence network for each fault.

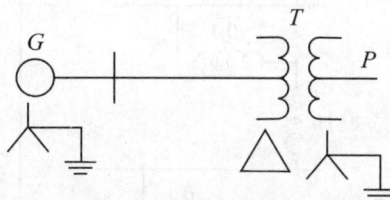

Figure 8.10   One-line diagram for Example 8.6.

Solution

$$S_b = 80 \text{ MVA}, \qquad V_b = 13.2 \text{ kV on generator side}$$

$$= 13.2 \times \frac{120}{13.2} = 120 \text{ kV at fault point } P$$

$$X_{1G} = X_{2G} = 0.1 \times \frac{80}{50} = 0.16 \text{ pu}$$

$$X_{0G} = 0.05 \times 80/50 = 0.08 \text{ pu}$$

$$X_{1T} = X_{2T} = X_{0T} = 0.1 \text{ pu}$$

(i) Three-phase fault   It is a symmetrical fault. Hence, only positive-sequence component is applicable. The connection diagram is shown in Figure 8.11.

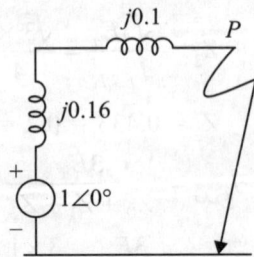

Figure 8.11 Connection diagram for three-phase fault of Example 8.6.

$$I_f = \frac{1\angle 0°}{j0.26} = 3.8462\angle -90° \text{ pu}$$

$$I_{\text{base}} = \frac{80}{\sqrt{3} \times 120} = 0.3849 \text{ kA}$$

$$I_{f\text{ actual}} = 3.8462 \times 0.3849 = 1.48 \text{ kA}$$

(ii) LG fault All the sequence networks are connected in series as shown in Figure 8.12.

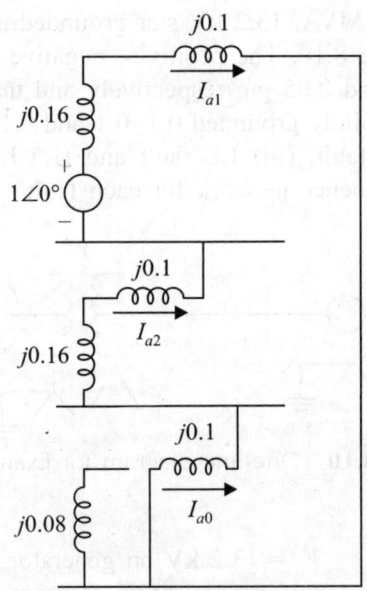

Figure 8.12 Connection diagram for LG fault of Example 8.6.

$$I_{a1} = \frac{1\angle 0°}{j0.26 + j0.26 + j0.1}$$

and

$$I_f = \frac{3 \times 1\angle 0°}{j0.26 + j0.26 + j0.1} = 4.8387\angle -90° \text{ pu}$$

$$I_{f\text{ actual}} = 4.8387 \times 0.3849 = 1.862 \text{ kA}$$

(iii) LL fault Positive- and negative-sequence networks are connected in parallel as shown in Figure 8.13.

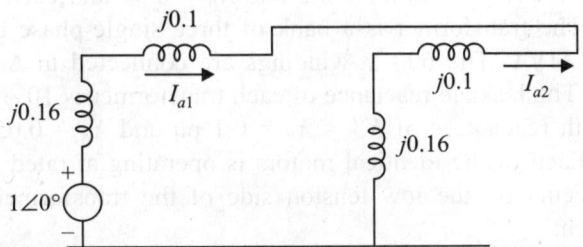

Figure 8.13 Connection diagram for LL fault of Example 8.6.

$$I_{a1} = \frac{1\angle 0°}{j0.26 + j0.26} \qquad I_b = \sqrt{3}\, I_{a1}$$

$$I_f = I_b = \frac{\sqrt{3} \times 1\angle 0°}{j0.26 + j0.26} = 3.3309 \ \angle{-90°}$$

$$I_{f\text{actual}} = 3.3309 \times 0.3849 = 1.282 \text{ kA}$$

(iv) LLG fault All the three sequence networks are connected in parallel as shown in Figure 8.14.

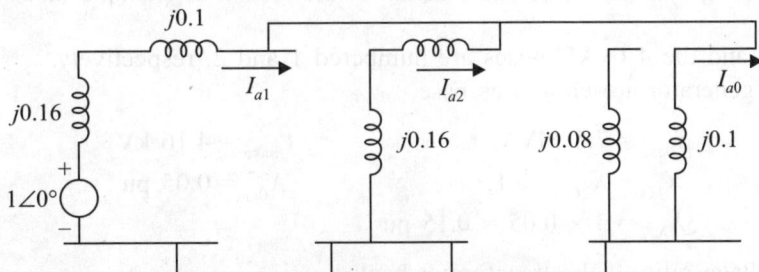

Figure 8.14 Connection diagram for LLG fault of Example 8.6.

$$I_{a1} = \frac{E_a}{Z_1 + \dfrac{Z_2 Z_0}{Z_2 + Z_0}} = \frac{1\angle 0°}{j0.26 + \dfrac{j0.26 \times j0.1}{j0.26 + j0.1}} = -j3.01$$

$$I_{a0} = -I_{a1}\frac{Z_2}{Z_2 + Z_0} = -2.1739 \text{ pu}$$

$$I_f = 3I_{a0} = -6.5217 \text{ pu}$$

$$I_{f\text{ actual}} = 6.5217 \times 0.3849 = 2.51 \text{ kA}$$

EXAMPLE 8.7 A group of four identical synchronous motors is connected through a transformer to a 4.16 kV bus of a generating plant. The motors are rated 600 V and operate at 89.5% efficiency when carrying full load at unity power factor and rated voltage. The sum of their output ratings is 4.476 MW. The reactances in per unit of each motor based on its own input MVA rating are $X'' = 0.2$, $X_2 = 0.2$ and $X_0 = 0.04$ and each is grounded through a reactance of 0.02 pu. The transformer is a bank of three single-phase units, each of which is rated 2400/600 V, 2.5 MVA. The 600 V windings are connected in Δ to the motors and the 2400 V windings in Y. The leakage reactance of each transformer is 10%. The generator is rated 7.5 MVA, 4.16 kV with reactances of $X'' = X_2 = 0.1$ pu and $X_0 = 0.05$ and X_n from neutral-to-ground is 0.05 pu. Each of the identical motors is operating at rated voltage, when a single line-to-ground fault occurs on the low tension side of the transformer bank. Determine the subtransient fault current.

Solution The one-line diagram of the system is shown in Figure 8.15.

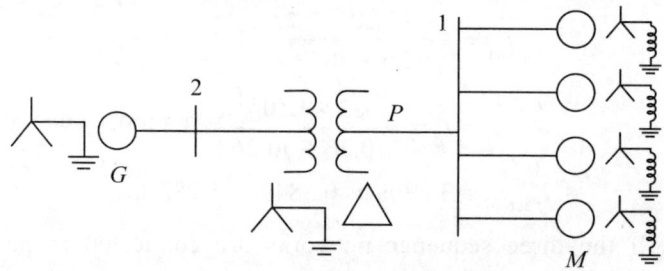

Figure 8.15 One-line diagram of the system of Example 8.7.

The 600 V and the 4.16 kV buses are numbered 1 and 2, respectively.
Ratings of generator are chosen as base.

$$S_{\text{base}} = 7.5 \text{ MVA}, \qquad V_{\text{base}} = 4.16 \text{ kV}$$
$$X_{1G} = X_{2G} = 0.1, \qquad X_{0G} = 0.05 \text{ pu}$$
$$3X_{NG} = 3 \times 0.05 = 0.15 \text{ pu}$$

Line-to-line voltage ratio of the transformer bank is

$$\frac{2400\sqrt{3}}{0.6} = \frac{4.16}{0.6} \text{ kV}$$

VA rating of transformer bank is

$$3 \times 2.5 = 7.5 \text{ MVA}$$

Treating group of motors as a single equivalent motor, the input rating is

$$\frac{4.476}{0.895 \times 1} = 5 \text{ MVA}$$

The reactance of the equivalent motor in per unit is the same on the base of the combined rating as the reactance of the individual motors on the base of the individual motor rating.
Since

$$\left(X \frac{nS_M}{S_M} \frac{1}{n} \right) = X$$

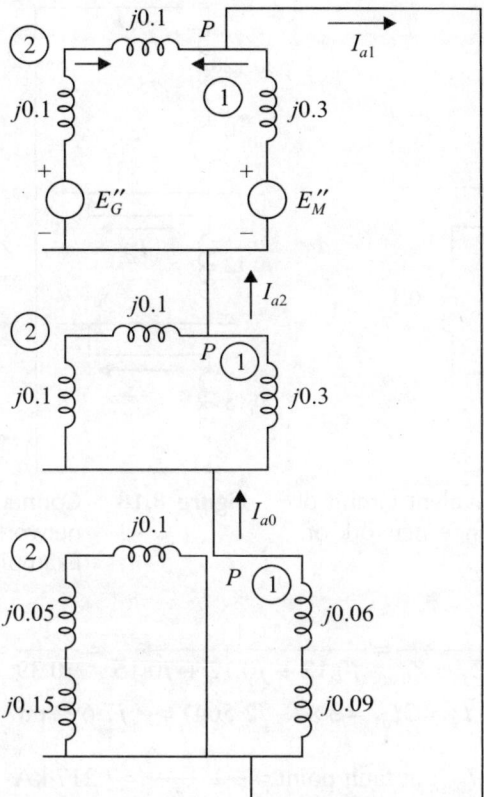

Figure 8.16 Connection of sequence networks for Example 8.7.

$$X_{1M} = 0.2 \times \frac{7.5}{5} = 0.3 \text{ pu}, \quad X_{2M} = 0.2 \times \frac{7.5}{5} = 0.3 \text{ pu}$$

$$X_{0M} = 0.04 \times \frac{7.5}{5} = 0.06 \text{ pu}$$

The neutral reactance in the zero-sequence network is

$$3X_{NM} = 3 \times 0.02 \times \frac{7.5}{5} = 0.09 \text{ pu}$$

Figure 8.16 shows the connection of the sequence networks.

Since the motors are operating at rated voltage, the pre-fault voltage of phase a at the fault is

$$V_f = 1 \text{ pu}$$

Neglecting pre-fault current,

$$E_a = E_G'' = E_M'' = V_f = 1\angle 0°$$

Thevenin's equivalent circuit of the positive-sequence network is shown in Figure 8.17.

After network reduction, the connection of sequence network is shown in Figure 8.18.

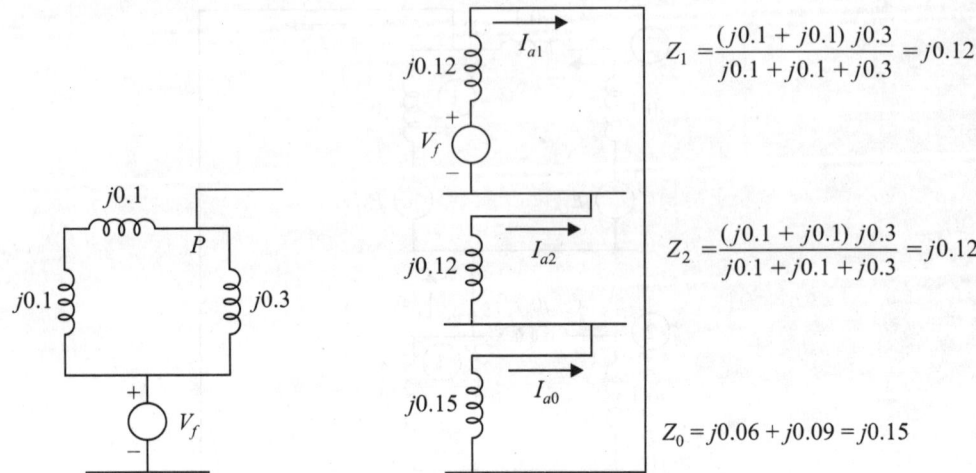

Figure 8.17 Thevenin equivalent circuit of positive-sequence network of Example 8.7.

Figure 8.18 Connection diagram of sequence networks after reduction for Example 8.7.

$$Z_1 = \frac{(j0.1 + j0.1)\, j0.3}{j0.1 + j0.1 + j0.3} = j0.12$$

$$Z_2 = \frac{(j0.1 + j0.1)\, j0.3}{j0.1 + j0.1 + j0.3} = j0.12$$

$$Z_0 = j0.06 + j0.09 = j0.15$$

$$I_{a1} = \frac{V_f}{Z_1 + Z_2 + Z_0} = \frac{1}{j0.12 + j0.12 + j0.15} = \frac{1}{j0.39} = -j2.564$$

$$I_f = 3I_{a1} = 3 \times (-j2.564) = -j7.692 \text{ pu}$$

$$I_{base} \text{ at fault point} = \frac{7.5}{\sqrt{3} \times 0.6} = 7.217 \text{ kA}$$

$$I_{f\,actual} = 7.692 \times 7.217 = 55.5 \text{ kA}$$

8.5 Fault through Impedance

The fault may not be a solid or bolted one, in which case the fault impedance may be non-zero and finite. It may be due to the impedance of the material causing fault and it may even be resistance of earth which is significant if the earth is dry. The effect of impedance in the fault is found by deriving equations similar to those for faults through zero impedance.

A balanced system remains symmetrical after the occurrence of a three phase fault having the same impedance between each line and a common point. Only positive-sequence currents flow. With the fault impedance z_f equal in all phases the voltage at the fault is $V_a = I_a z_f$ as shown in Figure 8.19(a).

Since only positive-sequence currents flow,

$$V_{a1} = I_{a1}z_f = V_f - I_{a1}Z_1$$

$$I_{a1} = \frac{V_f}{Z_1 + z_f} \tag{8.14}$$

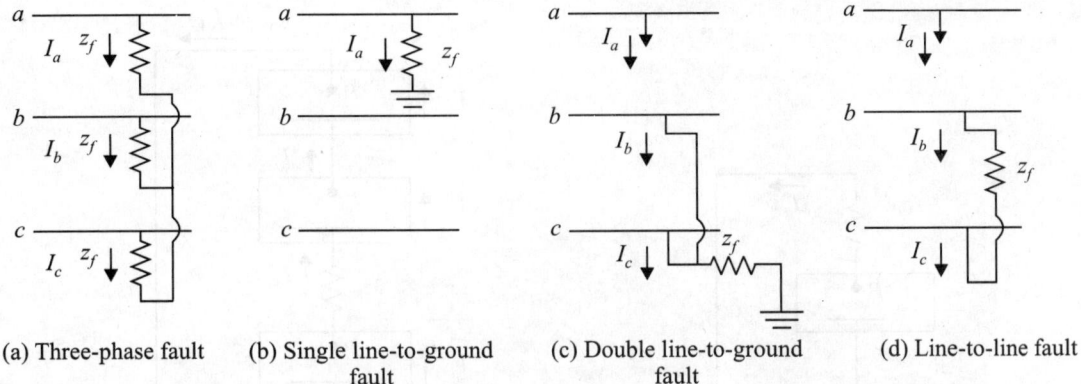

(a) Three-phase fault   (b) Single line-to-ground fault   (c) Double line-to-ground fault   (d) Line-to-line fault

Figure 8.19 Connection diagram of hypothetical stubs for various faults through impedance.

For an unloaded generator with neutral grounded, a single or double line-to-ground faults through z_f is no different with respect to the value of the fault current than the same type of fault without impedance but with z_f placed in the connection between the generator neutral and ground.

To account for impedance z_f in the neutral of a generator we add $3z_f$ to the zero-sequence networks. Thevenin's theorem enables us to apply the same reasoning to these types of faults on a power system. So the sequence network connections for a single line-to-ground fault and for a double line-to-ground fault are shown in Figure 8.19(b) and (c), respectively.

From Figure 8.19(b) for a single line-to-ground fault through z_f,

$$I_{a0} = I_{a1} = I_{a2}$$

$$I_{a1} = \frac{V_f}{Z_0 + Z_1 + Z_2 + 3z_f} \tag{8.15}$$

From Figure 8.20(d) for a double line-to-ground fault through z_f,

$$V_{a1} = V_{a2}$$

$$I_{a1} = \frac{V_f}{Z_1 + \dfrac{Z_2(Z_0 + 3z_f)}{Z_2 + Z_0 + 3z_f}} \tag{8.16}$$

A line-to-line fault through impedance is shown in Figure 8.19(d). The conditions at the fault are

$$I_a = 0, \qquad I_c = -I_b, \qquad V_c = V_b - I_b z_f$$

With $I_c = -I_b$ and $I_a = 0$, the symmetrical components of current are

$$I^{012} = A^{-1}I^{abc}$$

$$\begin{pmatrix} I_{a0} \\ I_{a1} \\ I_{a2} \end{pmatrix} = \frac{1}{3} \begin{pmatrix} 1 & 1 & 1 \\ 1 & a & a^2 \\ 1 & a^2 & a \end{pmatrix} \begin{pmatrix} 0 \\ I_b \\ -I_b \end{pmatrix}$$

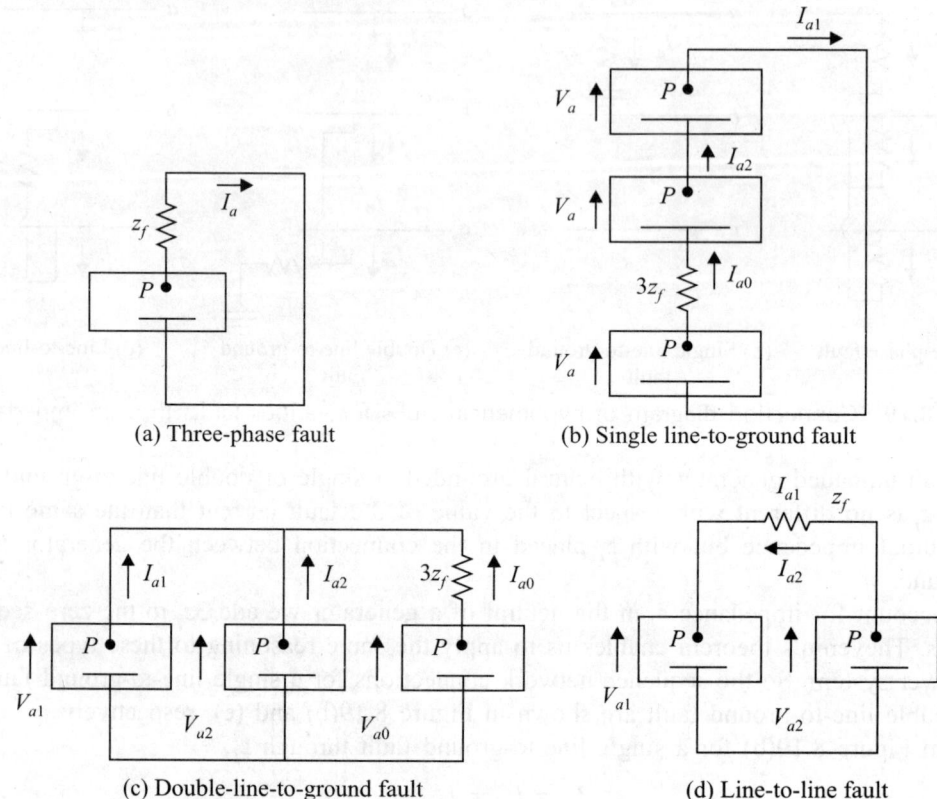

(a) Three-phase fault   (b) Single line-to-ground fault

(c) Double-line-to-ground fault   (d) Line-to-line fault

Figure 8.20 Connections of the sequence networks to simulate various types of faults through impedance at point P.

From this

$$I_{a0} = 0$$

$$I_{a1} = \frac{1}{3}(aI_b - a^2I_b) = \frac{1}{3}(a - a^2)I_b$$

$$I_{a2} = \frac{1}{3}(a^2I_b - aI_b) = \frac{1}{3}(a^2 - a)I_b$$

$$I_{a1} = -I_{a2}$$

The sequence components of voltage are given by

$$\begin{pmatrix} V_{a0} \\ V_{a1} \\ V_{a2} \end{pmatrix} = \frac{1}{3}\begin{pmatrix} 1 & 1 & 1 \\ 1 & a & a^2 \\ 1 & a^2 & a \end{pmatrix}\begin{pmatrix} V_a \\ V_b \\ V_b - I_b z_f \end{pmatrix}$$

$$3V_{a1} = V_a + (a + a^2)V_b - a^2I_b z_f$$

$$3V_{a2} = V_a + (a + a^2)V_b - aI_b z_f$$

$$3(V_{a1} - V_{a2}) = (a - a^2)\, I_b z_f = j\sqrt{3}\; I_b z_f \tag{8.17}$$

Since

$$I_{a1} = -I_{a2} \tag{8.18}$$

$$I_f = I_b = a^2 I_{a1} + a I_{a2} = (a^2 - a)I_{a1} = -j\sqrt{3}\, I_{a1} \tag{8.19}$$

$$\therefore \quad V_{a1} - V_{a2} = I_{a1} z_f \tag{8.20}$$

Since $I_{a0} = 0$, zero-sequence network is not present. From Eq. (8.18) and (8.19), the connection diagram for line-to-line fault is shown in Figure 8.23(c), inserting z_f between the fault points in the positive- and negative-sequence networks.

$$I_{a1} = \frac{V_f}{Z_1 + Z_2 + z_f} \tag{8.21}$$

$$I_f = \frac{-j\sqrt{3}V_f}{Z_1 + Z_2 + z_f} \tag{8.22}$$

where V_f is the pre-fault voltage in the positive-sequence network.

Bus impedance matrix can also be used to find Z_1, Z_2 and Z_0.

8.6 Unbalanced Fault Analysis using Bus Impedance Matrix

The bus impedance matrix, composed of positive-sequence impedances, is used to determine currents and voltages upon the occurrence of a balanced three-phase fault.

This method can easily be extended to unbalanced faults by realizing that the negative- and zero-sequence networks, and can be represented by bus impedance equivalent networks similar to the positive-sequence network. Figure 8.21 shows the sequence networks represented by the bus impedance equivalent networks (transfer impedances have, however, not been shown for clarity). The additional subscripts 1, 2 and 0 have been attached to the impedance to identify the sequence networks to which they belong. Depending upon the bus under fault and the type of fault, the connection diagram is drawn. For example, the connection diagram for LLG fault at bus n is shown in Figure 8.22.

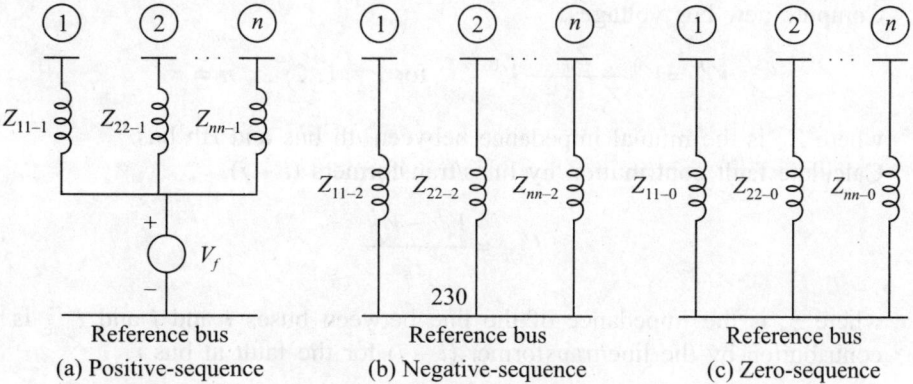

(a) Positive-sequence   (b) Negative-sequence   (c) Zero-sequence

Figure 8.21   Bus impedance equivalent sequence networks.

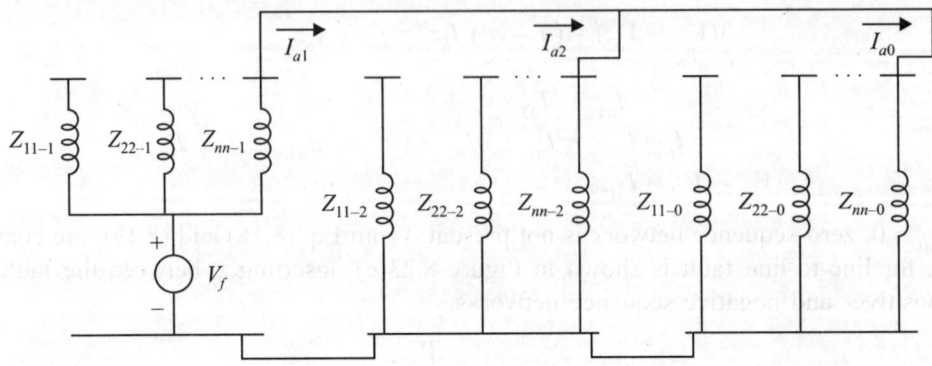

Figure 8.22 Connection diagram for LLG fault of bus impedance equivalent sequence networks.

Z_{nn-1}, Z_{nn-2} and Z_{nn-0} are equal to Z_1, Z_2 and Z_0 if the fault is on bus n. Thus self impedances are used to calculate the fault current and transfer impedances to calculate the un-faulted bus voltages. Contribution of fault current can be computed by dividing the potential difference across the line by its impedance, and adding it to pre-fault load current we can get post-fault current in various parts of the network.

Algorithm for calculation of unsymmetrical fault current using Z_{bus}

Step 1   Read the system data, fault impedance.

Step 2   Draw the reactance diagram.

Step 3   Form the bus impedance matrix Z_{bus}.

Step 4   Obtain pre-fault bus voltages V_r^0 and currents I_{ij}^0 over line i–j from power flow solution.

Step 5   Compute the fault current and bus voltage at rth bus.

$$I_{rf}'' = \frac{V_r^0}{Z_{rr} + z_f} \qquad \text{for } r = 1, 2, ..., n$$

where

z_f = fault impedance

Z_{rr} = self impedance of rth bus $V_r^f = z_f I_{rf}''$

Step 6   Compute new bus voltages.

$$V_i^f = V_i^0 \frac{Z_{ir}}{Z_{rr} + z_f} V_r^0 \qquad \text{for } i = 1, 2, ..., n \neq r$$

where Z_{ir} is the mutual impedance between ith bus and rth bus.

Step 7   Calculate fault contribution by lines/transformers $(i - j)$.

$$I_{ijrfc}'' = \frac{V_i^f - V_j^f}{z_{ij}}$$

where z_{ij} is the impedance of the line between buses i and j and I_{ijrfc} is the fault contribution by the line/transformer $(i - j)$ for the fault at bus r.

Step 8   Total fault current is obtained by superposing this current over the pre-fault current caused by all generators with subtransient reactance.

$$I''_{ijf} = I''_{ijrfc} + I^0_{ij}$$

Step 9   Check whether fault at all the buses are computed. If not, increase bus count $r = r + 1$ and go to Step 5.

Flow chart for calculation of unsymmetrical fault current using Z_{bus} is given in Figure 8.23.

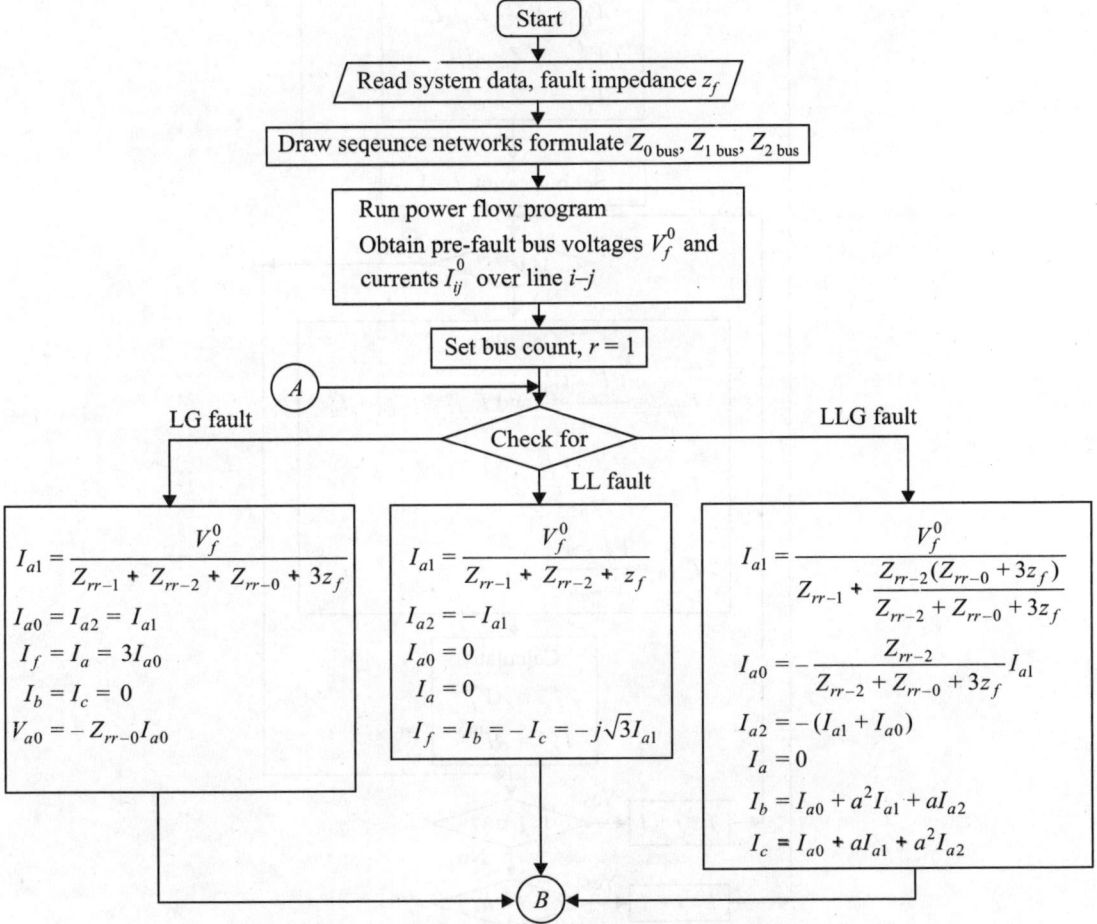

Figure 8.23   (*Contd.*)

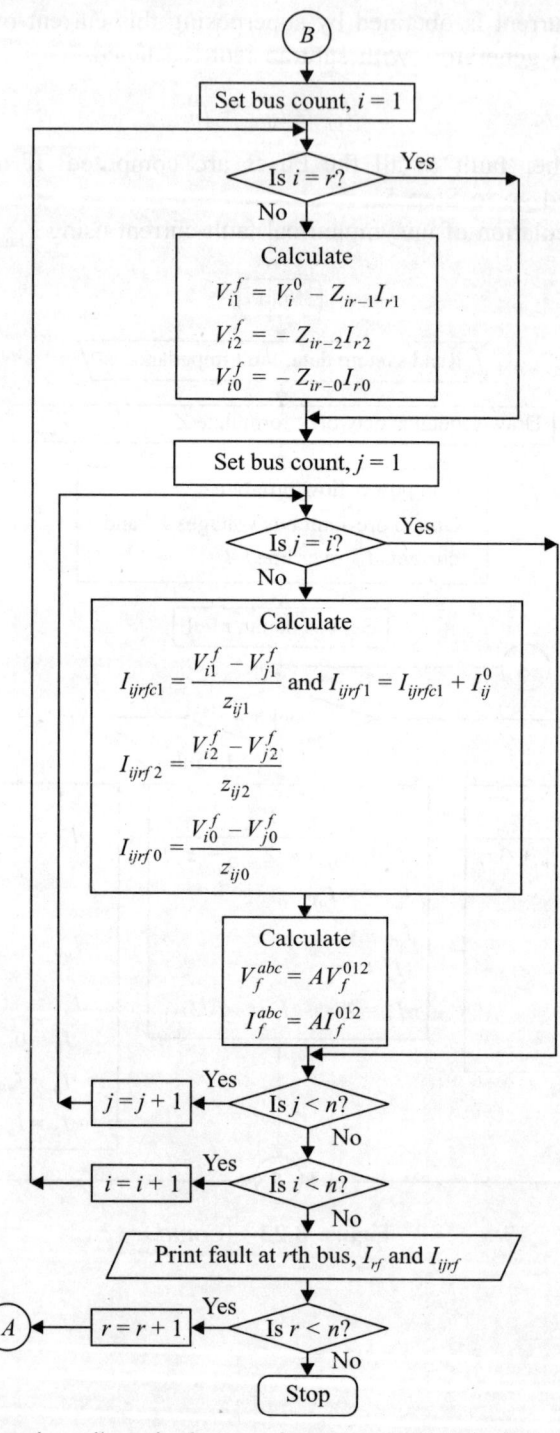

Figure 8.23 Flow chart for calculation of unsymmetrical fault current using Z_{bus}.

EXAMPLE 8.8   For the system given in Example 8.7, find the subtransient current in a single line-to-ground fault on bus 1 and also find the voltages-to-neutral at bus 2.

Solution   We refer to Figure 8.16 to find the bus impedance matrices of the three sequence networks.

Positive-sequence bus impedance matrix:

1.   Add $z_{10-1} = j\,0.3$ from bus 1 (new) to bus 0 (ref)

$$Z_{bus-1} = \begin{array}{c} \\ 1 \end{array} \begin{array}{c} 1 \\ (j0.3) \end{array}$$

2.   Add $z_{12} = j0.1$ from bus 2 (new) to bus 1 (existing)

$$Z_{bus-1} = \begin{array}{c} 1 \\ 2 \end{array} \begin{pmatrix} \overset{1}{j0.3} & \overset{2}{j0.3} \\ j0.3 & j0.3 + j0.1 \end{pmatrix}$$

$$= \begin{pmatrix} j0.3 & j0.3 \\ j0.3 & j0.4 \end{pmatrix}$$

3.   Add $z_{20} = j\,0.1$ from bus 2 (existing) to bus 0 (ref)

$$Z_{bus-1} = \begin{pmatrix} j0.3 & j0.3 \\ j0.3 & j0.4 \end{pmatrix} - \frac{1}{j0.4 + j0.1}\begin{pmatrix} j0.3 \\ j0.4 \end{pmatrix}(j0.3 \quad j0.4)$$

$$= \begin{pmatrix} j0.3 & j0.3 \\ j0.3 & j0.4 \end{pmatrix} - \begin{pmatrix} j0.18 & j0.24 \\ j0.24 & j0.32 \end{pmatrix}$$

$$= \begin{array}{c} 1 \\ 2 \end{array} \begin{pmatrix} \overset{1}{j0.12} & \overset{2}{j0.06} \\ j0.06 & j0.08 \end{pmatrix}$$

Similarly, negative-sequence bus impedance matrix is

$$Z_{bus-2} = \begin{array}{c} 1 \\ 2 \end{array} \begin{pmatrix} \overset{1}{j0.12} & \overset{2}{j0.06} \\ j0.06 & j0.08 \end{pmatrix}$$

Zero-sequence bus impedance matrix:

1.   Add $z_{10} = j0.15$ from bus 1 (new) to bus 0 (ref).

$$Z_{bus-0} = \begin{array}{c} \\ 1 \end{array} \begin{array}{c} 1 \\ (j0.15) \end{array}$$

2.   Add $z_{20} = \dfrac{j0.1 \times j0.2}{j0.1 + j0.2} = j0.067$ from bus 2 (new) to bus 0 (ref).

$$Z_{bus-0} = \begin{array}{c} 1 \\ 2 \end{array} \begin{pmatrix} \overset{1}{j0.15} & \overset{2}{0} \\ 0 & j0.067 \end{pmatrix}$$

The current in the fault on bus 1 is

$$I_{a1} = \frac{1}{j0.12 + j0.12 + j0.15} = -j2.564$$

$$I_f'' = 3I_{a1} = -j7.692$$

which agrees with the value found using network reduction in Example 8.7.

$$I_{f\,\text{actual}}'' = 7.692 \times 7.217 = 55.5 \text{ kA}$$

as

$$I_{\text{base}} = \frac{7.5}{\sqrt{3} \times 0.1} = 7.217 \text{ kA}$$

Using the transfer impedances the symmetrical components of voltage at bus 2 are

$$V_{a1} = V_f - I_{a1}Z_{21\text{-}1} = 1 - (-j2.564)\,(j0.06) = 0.8462$$

$$V_{a2} = -I_{a2}Z_{21\text{-}2} = -(-j2.564)\,(j0.06) = -0.1538$$

$$V_{a0} = -I_{a0}Z_{21\text{-}0} = 0$$

$$V_a = V_{a0} + V_{a1} + V_{a2} = 0 + 0.8463 - 0.1538 = 0.6924$$

$$V_b = V_{b0} + V_{b1} + V_{b2} = V_{a0} + a^2V_{a1} + aV_{a2}$$
$$= 0 + 0.8462(-0.5 - j0.866) - 0.1538(-0.5 + j0.866)$$
$$= -0.3462 - j0.8660$$

$$V_c = V_{c0} + V_{c1} + V_{c2} = V_{a0} + aV_{a1} + a^2V_{a2}$$
$$= 0 + 0.8462(-0.5 - j0.866) - 0.1538(-0.5 + j0.866)$$
$$= -0.3462 + j0.8660$$

Short Answer Questions

1 Draw the equivalent sequence network for a line-line bolted fault in a power system (Figure 8.24).

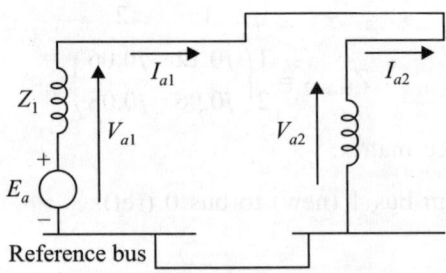

Figure 8.24 Equivalent sequence network for line-to-line bolted fault.

2. Name the fault in which all the sequence currents are equal.
   Single line-to-ground fault.

3. Name the faults which do not have zero-sequence current component.
   The faults are line-to-line fault and three-phase fault.

4. Name the fault in which positive- and negative-sequence currents together is equal to zero-sequence current in magnitude.

The fault is line-to-line fault or double line-to-ground fault.

5. Name the fault in which positive- and negative-sequence currents are equal in magnitude but opposite in direction.

The fault is line-to-line fault.

6. Name the fault in which positive- and negative-sequence voltages are equal.

The fault is double line-to-ground fault.

7. Name the faults involving ground.

The faults are:
- Single line-to-ground fault
- Double line-to-ground fault and
- Three-phase line-to-ground fault.

8. Name the various unsymmetrical faults in a power system.

The faults are:
- Single line-to-ground fault
- Line-to-line fault
- Double line-to-ground fault and
- Open conductor fault

9. State the boundary conditions for single line-to-ground fault.

The conditions for an single line-to-ground fault on phase a are

$$V_a = I_a z_f \text{ for fault through fault impedance } z_f$$
$$= 0 \text{ for solid fault and}$$
$$I_b = I_c = 0$$

Exercises

8.1 A generator having a solidly grounded neutral and rated 50 MVA, 30 kV has positive-, negative- and zero-sequence reactances of 25, 15 and 5 per cent respectively. What reactance must be placed in the generator neutral to limit the fault current for a bolted line-to-ground fault to that for a bolted three-phase fault?   [**Ans.** 1.8 Ω]

8.2 What reactance must be placed in the neutral of the generator of Problem 8.1 to limit the magnitude of the fault current for a bolted double line-to-ground fault to that for a bolted three-phase fault?   [**Ans.** 0.825 Ω]

8.3 Three 15 MVA, 30 kV synchronous generators A, B and C are connected via three reactors to a common bus bar as shown in Figure E8.1.

The neutral of generators A and B are solidly grounded, and the neutral of generator C is grounded through a reactor of 2 Ω. The generator data are tabulated in Table E8.1.

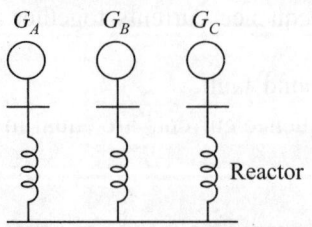

Figure E8.1 Circuit for Problem 8.3.

Table E8.1 Generation data

Item	X_1	X_2	X_0	Unit
G_A	0.25	0.155	0.056	pu
G_B	0.2	0.155	0.056	pu
G_C	0.2	0.15	0.06	pu
Reactor	6	6	6	Ω

A line-to-ground fault occurs on phase a of the common bus bar. Neglect pre-fault currents and assume generators are operating at their rated voltage. Determine
  (i) fault current in phase a.
  (ii) bolted line-to-line fault between phases b and c.
  (iii) double line-to-ground fault on phases b and c.

[**Ans.** (i) $12\angle-90°$ pu, (ii) -9.116 pu, (iii) $12.5\angle0°$ pu]

8.4 The reactance data for the power system shown in Figure E8.2 in per unit on a common base is given in Table E8.2

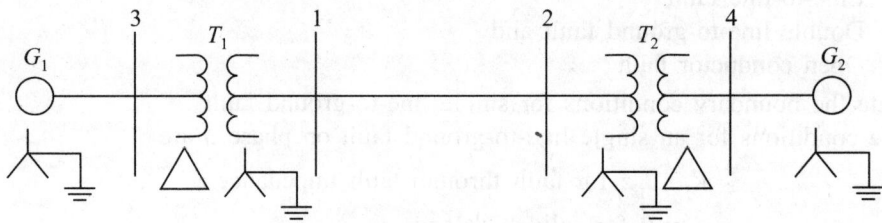

Figure E8.2 One-line diagram for Problem 8.4.

Table E8.2 Reactance data

Item	X_1	X_2	X_0
G_1	0.1	0.1	0.05
G_2	0.1	0.1	0.05
T_1	0.25	0.25	0.25
T_2	0.25	0.25	0.25
Line 1–2	0.3	0.3	0.5

Obtain the Thevenin's sequence impedances for the fault at bus 1 and compute the fault current in per unit for the following faults:
  (i) A bolted three-phase fault   (ii) A bolted single line-to-ground fault
  (iii) A bolted line-to-line fault   (iv) A bolted double line-to-ground fault

[**Ans.** $4.395\angle-90°$, $4.669\angle-90°$, $-3.807\angle-90°$, $4.979\angle90°$ pu]

8.5 The zero-, positive- and negative-sequence bus impedance matrices for a three-bus power system are

$$Z_{bus}^0 = j\begin{pmatrix} 0.2 & 0.05 & 0.12 \\ 0.05 & 0.1 & 0.08 \\ 0.12 & 0.08 & 0.3 \end{pmatrix} \text{pu} \qquad Z_{bus}^1 = Z_{bus}^2 = j\begin{pmatrix} 0.16 & 0.1 & 0.15 \\ 0.10 & 0.2 & 0.12 \\ 0.15 & 0.12 & 0.25 \end{pmatrix} \text{pu}$$

Determine the per-unit fault current and the bus voltages during fault at bus 2 for
- (i) A bolted three-phase fault
- (ii) A bolted single line-to-ground fault
- (iii) A bolted line-to-line fault
- (iv) A bolted double line-to-ground fault

[**Ans.** $5\angle-90°$, $6\angle-90°$, $-4.33\angle-90°$, $7.5\angle90°$ pu]

8.6 Determine the fault current and MVA at faulted bus for a line-to-ground (solid) fault at bus 4 of a system as shown in Figure E8.3.

G_1, G_2 : 100 MVA, 11 kV, $X^+ = X^- = 15\%$, $X° = 5\%$, $X_n = 6\%$

T_1, T_2 : 100 MVA, 11/220, kV, $X_{leak} = 9\%$

L_1, L_2 : $X^+ = X^- = 10\%$, $X° = 10\%$ on a base of 100 MVA.

Consider a fault on phase a.   [**Ans.** 42.75 kA, 814.55 MVA]

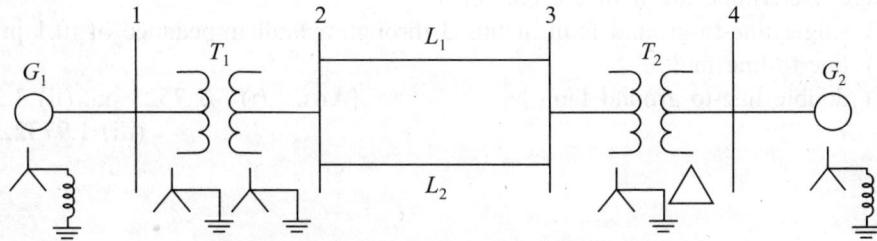

Figure E8.3 Circuit for Problem 8.6.

8.7 The one-line diagram of a simple power system is shown in Figure E8.4. The system data expressed in per unit on a common 100 MVA base is tabulated in Table E8.3.

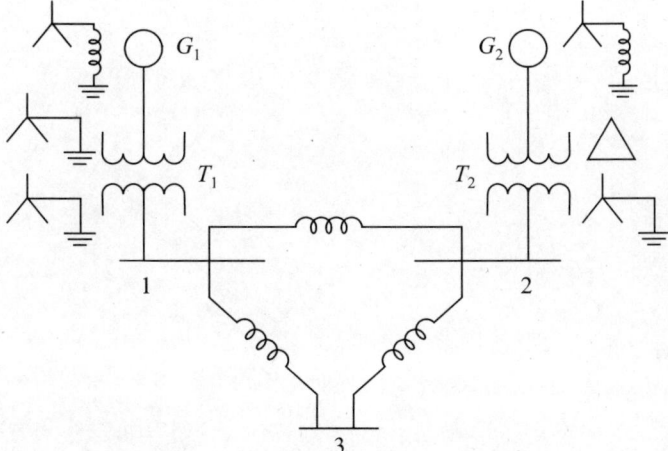

Figure E8.4 One-line diagram for Problem 8.7.

Table E8.3 System data

Item	Voltage, kV	X_1	X_2	X_0
G_1	20	0.15	0.15	0.05
G_2	20	0.15	0.15	0.05
T_1	20/220	0.1	0.1	0.1
T_2	20/220	0.1	0.1	0.1
L_{12}	220	0.125	0.125	0.3
L_{13}	220	0.15	0.15	0.35
L_{23}	220	0.25	0.25	0.7125

The neutral of each generator is grounded through a current limiting reactor of 0.25/3 pu on a 100 MVA base. The generators are running on no load at their rated voltage. Determine the fault current for a

(i) single line-to-ground fault at bus 3 through a fault impedance of $j0.1$ pu.
(ii) line-to-line fault.
(iii) double line-to-ground fault.   [**Ans.** (i) $-j2.7523$ pu, (ii) 3.2075 pu,

(iii) $1.9732\angle90°$ pu]

CHAPTER 9

Power System Stability

A power system consists of a number of synchronous machines operating in synchronism under all conditions. When the system is subjected to some form of disturbance, there is a tendency for the system to develop restoring force to bring it to normal or stable condition. The ability of a system to reach a normal or stable condition after being disturbed is called **stability**. Stability in a power system is defined as that attribute of the system, which enables it to develop restoring forces between the elements thereof to counter the disturbing forces so as to restore the state of equilibrium between the elements. Stability depends on both the initial operating state of the system and the severity of the disturbance.

Stability studies are helpful in determining the power transfer capability between power systems and in designing the protection and relaying system and parameters of control devices (excitation system, automatic voltage regulator, power system stabilizer and governor).

The following are the causes of disturbances:

- Natural causes such as lightning resulting in flashover across insulators
- Inadvertent causes such as maloperation of protection
- Intended action such as opening/closing of circuit breakers by the system operator.

9.1 Classification of Stability

Power system stability can be classified into angle stability and voltage stability depending on the parameter affected, whether loss of synchronism is due to increase in power angle difference or voltage exceeding acceptable limit.

Power system stability can also be classified into mid-term and long-term stability based on the duration of study after the disturbance and the type of disturbance.

9.1.1 Rotor Angle Stability

Rotor angle stability is the ability of interconnected synchronous machines of a power system to remain in synchronism. The stability problem involves the study of the electro-mechanical oscillations involving the exchange of energy between network and generator mechanical system.

The rotor angle stability phenomenon can further be subdivided into two categories based on the severity/size of disturbance and the study period after disturbance. They are:

- Large signal or transient stability
- Small signal or steady-state

Transient stability

This is the ability of the system to return to a stable operating condition after large disturbances like occurrence of fault, sudden outage of a line or generator, sudden application or removal of large loads, etc. The resulting system response involves large excursions of generator rotor angles and is influenced by the nonlinear power angle relationship. Transient stability studies are generally based on a first swing rather than multiswing. The generator model is represented by a transient internal voltage E' behind transient reactance X'_d. Excitation and turbine-governor control are not included in this modelling. Multi-swing stability studies extend over a longer period. Excitation and turbine-governor controls are included in this modelling.

The rotor angle response to a transient disturbance, known as swing curve, is shown in Figure 9.1.

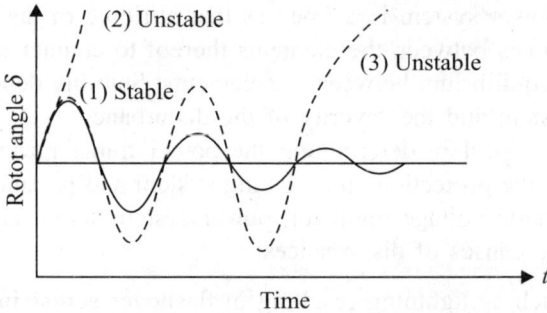

Figure 9.1 Rotor angle response to transient disturbance.

Case 1: The rotor angle increases to a maximum and then decreases and oscillates with decreasing amplitude until it reaches steady state and is the result of positive damping.

Case 2: The rotor angle continues to increase steadily until synchronism is lost. This form of instability is referred to as nonoscillatory instability or first swing instability and is caused by insufficient or negative synchronizing torque.

Case 3: The system is stable in the first swing but becomes unstable as a result of growing amplitude of oscillation due to negative damping and is referred to as oscillatory instability.

The study period is usually up to 3 to 5 seconds and it may extend to 10 seconds for very large system.

Steady-state stability

This is the ability of the power system to return to a stable operating condition or remain in synchronism after a small disturbance such as gradual variation in system variables such as rotor angle, voltage, etc.

Stability depends on both initial operating state of the system and severity of the disturbances. Steady-state stability is further subdivided into static and dynamic stability.

Static stability: It refers to inherent stability that prevails after small disturbance without the aid of automatic control devices.

Dynamic stability: It refers to the ability of the system to remain in synchronism when it is subjected to small disturbance like oscillatory one (usually following transient state), with the help of automatic control devices.

The oscillatory stability depends on damping and the system becomes stable if the rotor oscillations decrease in amplitude due to positive damping torque or unstable if the rotor oscillations increase in amplitude due to negative damping torque or unstable control action. *Case* 3 under large signal stability also refers to this type of stability namely dynamic stability.

The various modes of oscillation are:

(a) Local mode corresponding to swinging of units in a plant with respect to the rest of the system (f = 1 to 2 Hz).
(b) Inter-area mode corresponding to swinging of many machines in one area against those in others (f = 0.4 to 0.7 Hz).
(c) Control mode instability due to poor tuning of controllers, i.e. exciter (AVR), governor, HVDC converters and SVC.
(d) Torsional mode instability due to interaction between turbine-generator shaft system, electrical system with series capacitor compensation and HVDC control.

For pure angle stability of a synchronous generator, the power system considered for studies is composed of that generator connected to an infinite bus through its reactance as shown in Figure 9.2. This simplest system for analysis of angle stability is called Single Machine Infinite Bus (SMIB) system.

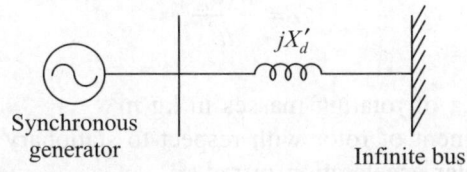

Figure 9.2   System for pure angle stability studies.

9.1.2   Voltage Stability

Voltage instability implies an uncontrolled decrease in voltage monotonically, triggered by a disturbance, leading to voltage collapse and is primarily caused by the characteristics and dynamics of the connected load.

Although the frequency may increase due to load reduction on voltage collapse, the generators normally remain in synchronism. For pure voltage stability, the power system considered for study is composed of load connected to an infinite bus through a line as shown in Figure 9.3.

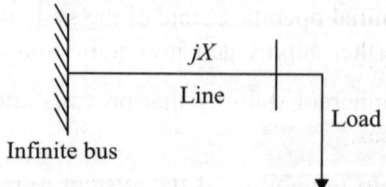

Figure 9.3   System for pure voltage stability study.

9.1.3 Mid-term Stability

Mid-term stability comes into play when severe disturbances affect the system causing large frequency and voltage excursions. It involves both fast and slow dynamics and the study period is up to several minutes.

9.1.4 Long-term Stability

Long-term stability is involved with slow dynamics and uniform frequency. The study period is up to several minutes.

9.2 Swing Equation

The accelerating torque on a rotating machine is given by

$$T_a = J\alpha \tag{9.1}$$
$$T_a = T_m - T_e \tag{9.2}$$

$$\alpha = \frac{d^2\theta_m}{dt^2} \tag{9.3}$$

where

J = moment of inertia of rotating masses in kg m^2
θ_m = angular displacement of rotor with respect to stationary axis, in mechanical radians
α = mechanical angular acceleration in rad/s^2
t = time in seconds
T_m = mechanical or shaft torque supplied by the prime-mover less retarding torque due to rotational losses, in Nm
T_e = the electrical or electromagnetic torque including losses in Nm
T_a = accelerating torque in Nm.

Figure 9.4 indicates the representation of a machine rotor.

Under steady-state, $T_e = T_m$ and therefore $T_a = 0$ and the rotor rotates at synchronous speed.

Since θ_m is measured with respect to a stationary reference axis on the stator, it is an absolute measure of rotor angle. Consequently, it continuously increases with time even at constant synchronous speed. Since the rotor speed relative to synchronous speed is required it is convenient to measure the rotor angular position with respect to a reference axis, which rotates at synchronous speed.

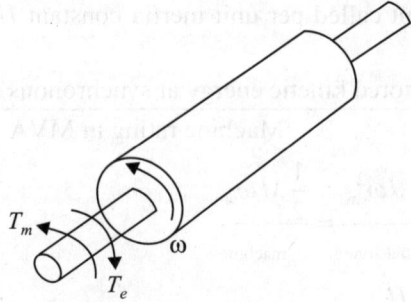

Figure 9.4 Representation of a machine rotor.

$$\delta_m = \theta_m - \omega_{sm}t \qquad (9.4)$$

where

ω_{sm} = synchronous speed of the machine in mechanical radians per second

δ_m = angular displacement of the rotor from the synchronously rotating reference axis, in mechanical radians

The derivative of Eq. (9.4) with respect to time is

$$\frac{d\delta_m}{dt} = \frac{d\theta_m}{dt} - \omega_{sm} \qquad (9.5)$$

and

$$\frac{d^2\delta_m}{dt^2} = \frac{d^2\theta_m}{dt^2} \quad (\text{since } \omega_{sm} \text{ is constant}) \qquad (9.6)$$

$$J\frac{d^2\delta_m}{dt^2} = T_m - T_e \qquad (9.7)$$

$$J\omega_m\frac{d^2\delta_m}{dt^2} = T_m\omega_m - T_e\omega_m = P_m - P_e = P_a \qquad (9.8)$$

where,

P_m = shaft power less rotational losses

P_e = electrical power crossing the air gap

P_a = accelerating power

Neglecting the rotational and armature copper losses,

P_m = power supplied by the prime mover

P_e = electrical power output

$$M\frac{d^2\delta_m}{dt^2} = P_m - P_e \qquad (9.9)$$

where

$M = J\omega_{sm}$ is the angular momentum of the rotor at synchronous speed ($\omega_m \approx \omega_{sm}$) and called *inertia constant* of the machine in MJ s/mech. rad.

Normally, another constant called per unit inertia constant H in seconds is used in stability studies. H is defined as

$$H = \frac{\text{Stored kinetic energy at synchronous speed in MJ}}{\text{Machine rating in MVA}}$$

$$= \frac{\frac{1}{2} J \omega_{sm}^2}{S_{machine}} = \frac{\frac{1}{2} M \omega_{sm}}{S_{machine}} \qquad (9.10)$$

or

$$M = \frac{2H}{\omega_{sm}} S_{machine} \qquad (9.11)$$

Note: $M = J\omega_{sm} = J\omega_s/p$ MJ s/mech. rad.   (p = pair of poles)

   $= J\omega_s/p^2$ MJ s/elect. rad.

   $= J\omega_s/p^2 \times \pi/180$ MJ s/elect. deg.

Substituting in Eq. (9.9), we get

$$\frac{2H}{\omega_{sm}} \frac{d^2\delta_m}{dt^2} = \frac{P_m}{S_{machine}} - \frac{P_e}{S_{machine}}$$

$$\frac{2H}{\omega_{sm}} \frac{d^2\delta_m}{dt^2} = P_m - P_e \text{ per-unit}$$

$$\frac{2H}{\omega_s} \frac{d^2\delta}{dt^2} = P_m - P_e \text{ per-unit} \qquad (9.12)$$

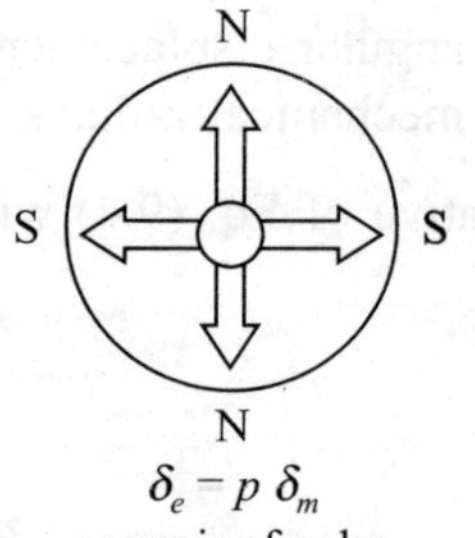

where

$$\delta_e = p \, \delta_m$$
$$p = \text{pair of poles}$$

Figure 9.5   Electrical vs mechanical angle.

δ and ω_s have consistent units which may be mechanical or electrical degrees or radians as shown in Figure 9.5

H and t have consistent units since MJ/MVA is in units of time in seconds

P_m and P_e must be in pu on the same base as H.

Equation (9.12) is called the swing equation of the machine. This second order differential equation can be written as two first order differential equations.

$$\frac{2H}{\omega_s} \frac{d\omega}{dt} = P_m - P_e \qquad (9.13)$$

$$\frac{d\delta}{dt} = \omega - \omega_s \qquad (9.14)$$

Equation (9.12) can also be written as

$$\frac{H}{\pi f} \frac{d^2\delta}{dt^2} = P_m - P_e \quad (\text{since } \omega_s = 2\pi f) \qquad (9.15)$$

where δ is in electrical radian.

$$\frac{H}{180 f} \frac{d^2\delta}{dt^2} = P_m - P_e \qquad (9.16)$$

when δ is in electrical degree.

In a power system, there may be many synchronous machines and it becomes necessary to convert the H constants of the machines to a common base MVA, say system MVA.

$$H_{\text{system}} = H_{\text{machine}} \times \frac{S_{\text{machine}}}{S_{\text{system}}} \qquad (9.17)$$

Depending upon the location of disturbance some of the machines called coherent machines will swing together and others incoherently. In fact, machines connected in parallel at a bus are coherent machines and they swing together. The machines connected in parallel but through their transfer impedances are noncoherent machines and they swing differently. The H constants of these machines can be combined into an equivalent H constant.

Coherent machines

Consider n coherent machines with H constants on common base MVA. For machine n

$$\frac{2H_n}{\omega_s} \frac{d^2 \delta_n}{dt^2} = P_{mn} - P_{en} \qquad (9.18)$$

Since the rotors swing together, $\delta_1 = \delta_2 = \cdots = \delta_n = \delta$.

Adding the n equations, we get

$$\frac{2H}{\omega_s} \frac{d^2 \delta}{dt^2} = P_m - P_e \qquad (9.19)$$

where

$$\left. \begin{array}{l} H = H_1 + H_2 + \cdots + H_n \\ P_m = P_{m1} + P_{m2} + \cdots + P_{mn} \\ P_e = P_{e1} + P_{e2} + \cdots + P_{en} \end{array} \right\} \qquad (9.20)$$

or

Energy stored by the equivalent machine should be equal to the sum of energy stored by individual machines.

$$S_b H_e = S_1 H_{11} + S_2 H_{22} + \cdots + S_n H_{nn} \text{ where, } H_{nn} \text{ is on its own MVA} \qquad (9.21)$$

$$H = \frac{S_1}{S_b} H_{11} + \frac{S_2}{S_b} H_{22} + \cdots + \frac{S_n}{S_b} H_{nn} \qquad (9.22)$$

$$= H_1 + H_2 + \cdots + H_n \text{ on common base MVA, } S_b.$$

Noncoherent machines

For any pair of noncoherent machines separated by transfer impedance the swing equations are combined to compute the relative angle between their rotors that is going to decide about their synchronous operation.

$$\frac{2}{\omega_s} \left(\frac{d^2 \delta_1}{dt^2} - \frac{d^2 \delta_2}{dt^2} \right) = \left(\frac{P_{m1} - P_{e1}}{H_1} - \frac{P_{m2} - P_{e2}}{H_2} \right) \qquad (9.23)$$

Multiplying both sides by $[H_1 H_2/(H_1 + H_2)]$ and rearranging, we get

$$\frac{2}{\omega_s}\left(\frac{H_1 H_2}{H_1 + H_2}\right)\left(\frac{d^2(\delta_1 - \delta_2)}{dt^2}\right) = \left(\frac{P_{m1}H_2 - P_{m2}H_1}{H_1 + H_2} - \frac{P_{e1}H_2 - P_{e2}H_1}{H_1 + H_2}\right) \tag{9.24}$$

Simplifying in the form of basic swing equation,

$$\frac{2H_{12}}{\omega_s}\frac{d^2\delta_{12}}{dt^2} = P_{m12} - P_{e12} \tag{9.25}$$

where

$$\delta_{12} = \delta_1 - \delta_2 \text{ is the relative angle} \tag{9.26}$$

$$H_{12} = \frac{H_1 H_2}{H_1 + H_2} \text{ is the equivalent inertia}$$

$$\frac{1}{H_{12}} = \frac{1}{H_1} + \frac{1}{H_2} \tag{9.27}$$

$$P_{m12} = \frac{P_{m1}H_2 - P_{m2}H_1}{H_1 + H_2} \text{ is the weighted input} \tag{9.28}$$

$$P_{e12} = \frac{P_{e1}H_2 - P_{e2}H_1}{H_1 + H_2} \text{ is the weighted output} \tag{9.29}$$

In a two-machine system having one generator and a synchronous motor connected by a network of pure reactance

$$P_{m1} = -P_{m2} = P_m$$
$$P_{e1} = -P_{e2} = P_e$$
$$\therefore \qquad P_{m12} = P_m \text{ and } P_{e12} = P_e$$

and the swing equation is

$$\frac{2H_{12}}{\omega_s}\frac{d^2\delta_{12}}{dt^2} = P_m - P_e \tag{9.30}$$

EXAMPLE 9.1 A 4 pole, 50 Hz, 60 MVA turbo generator has a moment of inertia of 9×10^3 kg m^2. Determine (i) the kinetic energy, (ii) inertia constants, (iii) the acceleration in deg. per s^2 and in rpm/s if the input power is 20 MW and the output power is 15 MW.

Solution

$$\text{Kinetic energy (KE)} = \frac{1}{2}J\omega_{sm}^2 = \frac{1}{2}J\left(\frac{\omega_s}{p}\right)^2 = \frac{1}{2} \times 9 \times 10^3 \left(2 \times \pi \times \frac{50}{2}\right)^2 \times 10^{-6} \text{ MJ}$$

$$= 111.03 \text{ MJ} \left(\text{since } p = \frac{4}{2} = 2\right)$$

$$H = \frac{KE}{S} = \frac{111.03}{60} = 1.851 \text{ MJ/MVA}$$

$$M = \frac{2HS}{\omega_s} = \frac{2 \times 1.851 \times 60}{2 \times \pi \times 50} \text{ MJ s/elec. rad.}$$

$$\frac{1.851 \times 60}{180 \times 50} = 0.01234 \text{ MJ s/elec. deg.}$$

or

$$M = J\omega_{sm} = \frac{J\omega_s}{p} \text{ MJ s/mech. rad.}$$

$$= \frac{J\omega_s}{p^2} \text{ MJ s/elec. rad.} = \left(\frac{J\omega_s}{p^2}\right)\left(\frac{\pi}{180}\right) \text{ MJ s/elec. deg.}$$

$$= \left(\frac{9 \times 10^3 \times 2\pi \times 50}{2^2}\right) \times \left(\frac{\pi}{180}\right) \times 10^{-6} = 0.01234 \text{ MJ s/elec. deg.}$$

or

$$M = J\omega_{sm} = \left(\frac{\frac{1}{2}J\omega_{sm}^2}{\frac{1}{2}\omega_{sm}}\right) = \frac{2 KE}{\omega_{sm}} \text{ MJ s/mech. rad.}$$

$$= \frac{2 KE}{\omega_s} \text{ MJ s/elec. rad.} = \frac{KE}{\pi f} \text{ MJ s/elec. rad.}$$

$$= \frac{KE}{180 f} \text{ MJ s/elec. deg.}$$

$$= \frac{111.03}{180 \times 50} = 0.01234 \text{ MJ s/elec. deg.}$$

$$M\alpha = P_a = P_m - P_e$$

$$0.01234 \, \alpha = 20 - 15 = 5$$

$$\alpha = \frac{5}{0.01234} = 405 \text{ elec. deg./s}^2$$

$$= \frac{405}{p} \text{ mech. deg./s}^2 = 405/2/360 \text{ rps}^2 \quad \text{since 1 rotation} = 360 \text{ mech. deg.}$$

$$= \frac{405 \times 60}{2 \times 360} \text{ rpm/s} = 33.77 \text{ rpm/s}$$

EXAMPLE 9.2 Two generating units operate in parallel within the same power plant and have the following ratings:

Unit 1:   500 MVA, 4.8 MJ/MVA
Unit 2:   1333 MVA, 3.27 MJ/MVA

Calculate the equivalent H-constant for the two units on a 100 MVA base.

Solution   Total KE $= S_1 H_{11} + S_2 H_{22}$

$$= 500 \times 4.8 + 1333 \times 3.27 = 6759 \text{ MJ}$$

H constant for the equivalent machine on 100 MVA base

$$= \frac{6759}{100} = 67.59 \text{ MJ/MVA}$$

or

Since the machines are connected to the same bus, they are coherent and swing together.

$$\therefore \qquad H = H_1 + H_2 \quad \text{on a common base.}$$

$$H_1 = 4.8 \times \frac{500}{100} = 24 \text{ MJ/MVA}$$

and

$$H_2 = 3.27 \times \frac{1333}{100} = 43.59 \text{ MJ/MVA}$$

$$\therefore \qquad H = 24 + 43.59 = 67.59 \text{ MJ/MVA}$$

9.3 Power Angle Equation

Two-machine problems of following two types are studied to illustrate the stability conditions:

(i) One machine of finite inertia swinging with respect to an infinite bus (constant internal voltage having zero impedance and infinite inertia)

(ii) Two finite inertia machines swinging with respect to each other.

Fundamental assumptions made in all stability studies are:

(a) Only synchronous frequency currents and voltages are considered. Consequently, dc offset currents and harmonic components are neglected.

(b) Generated voltage is considered unaffected by machine speed variations.

Figure 9.6 schematically represents a generator at bus 1 supplying power through a transmission system to a receiving end system at bus 2. The transmission system consists of linear passive components such as transformers, transmission lines, reactors and capacitors and includes the transient reactance of the generator at bus 1 and also that of the synchronous machine, either generator or motor, at bus 2 if it is not an infinite system. E_1' and E_2' represent their transient internal voltages. The bus admittance matrix of the transmission network is

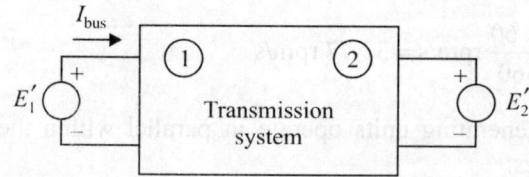

Figure 9.6   Schematic diagram of a power system.

$$Y_{\text{bus}} = \begin{pmatrix} Y_{11} & Y_{12} \\ Y_{21} & Y_{22} \end{pmatrix} \qquad (9.31)$$

We know power injected at bus p is given by

$$S_p = E_p I_p^* \quad \text{or} \quad S_p^* = E_p^* I_p \qquad (9.32)$$

$$I_p = \sum_{q=1}^{n} Y_{pq} E_q \qquad (9.33)$$

$$\therefore \qquad P_p - jQ_p = \sum_{q=1}^{n} E_p^* Y_{pq} E_q \qquad (9.34)$$

Equating real parts on both sides, we obtain

$$P_p = \sum_{q=1}^{n} E_p E_q Y_{pq} \cos \angle(\theta_{pq} + \delta_q - \delta_p) \qquad (9.35)$$

For $n = 2$, the real power injected at bus 1 is

$$P_1 = (E_1')^2 Y_{11} \cos \theta_{11} + E_1 E_2 Y_{12} \cos(\theta_{12} + \delta_2 - \delta_1)\sin \delta \qquad (9.36)$$

For purely reactive network,

$$\theta_{11} = \theta_{12} = 90°$$

Let

$$\delta_1 - \delta_2 = \delta \qquad (9.37)$$

As resistance has been neglected, there is no real power loss. Therefore

$$P_2 = P_1 = P = E_1' E_2' Y_{12} \sin \delta \qquad (9.38)$$

or

$$P = \frac{E_1' E_2'}{X} \sin \delta \qquad (9.39)$$

where X is the transfer reactance.

Equation (9.39) is called the *power angle equation* and its graph showing P as a function of δ is called the *power angle curve*.

The power angle Eq. (9.39) can also be derived from the phasor diagram shown in Figure 9.7.

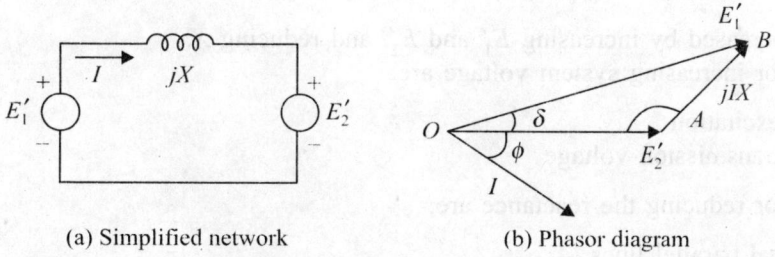

(a) Simplified network   (b) Phasor diagram

Figure 9.7 Phasor diagram for a power system with two buses.

From ΔOAB,

$$\frac{AB}{\sin \delta} = \frac{OB}{\sin(90 + \phi)} \qquad (9.40)$$

$$\frac{IX}{\sin \delta} = \frac{E_1'}{\cos \phi}$$

$$I \cos \phi = \frac{E_1'}{X} \sin \delta \qquad (9.41)$$

Multiplying both sides by E_2', we get

$$E_2'I \cos\phi = \frac{E_1'E_2'}{X} \sin\delta$$

where

ϕ = power factor angle
δ = power angle
E_1' = internal voltage of generator at bus 1
E_2' = internal voltage of synchronous machine at bus 2
X = reactance between the synchronous machines
I = current flow

As real power loss is negligible,

$$P_2 = P_1 = P = \frac{E_1'E_2'}{X} \sin\delta = P_{max} \sin\delta \qquad (9.42a)$$

where,

$$P_{max} = \frac{E_1'E_2'}{X} \qquad (9.42b)$$

9.3.1 Methods of Improving Steady-state Stability

Steady-state stability limit is the level of maximum power that can be transmitted without losing stability.

$$P_{max} = \frac{E_1'E_2'}{X}$$

P_{max} can be increased by increasing E_1' and E_2', and reducing X.
Methods for increasing system voltage are:

- Higher excitation
- Higher transmission voltage

Methods for reducing the reactance are:

- Additional parallel lines
- Use of bundled conductors
- Series compensation of lines
- Low machine and transformer impedances

EXAMPLE 9.3 Figure 9.8 shows a generator connected through two parallel transmission lines to a large metropolitan system considered as an infinite bus. The machine is delivering 1.0 pu power and both generator terminal voltage and infinite bus voltage are 1.0 pu. The diagram indicates the value of the reactances on a common base system. Determine the power angle equation for the system applicable to the operating condition.

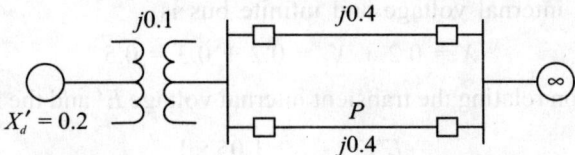

Figure 9.8   One-line diagram for Example 9.3.

Solution   The reactance diagram for the system is shown in Figure 9.9.

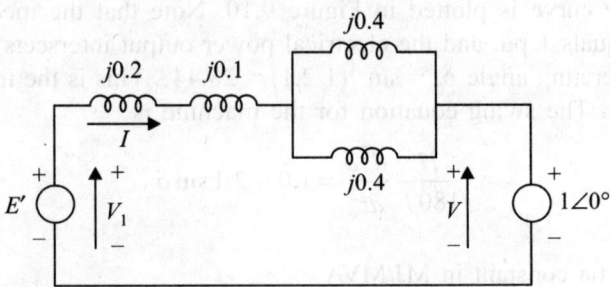

Figure 9.9   Reactance diagram for Example 9.3.

The series reactance between terminal voltage and the infinite bus is

$$X_1 = 0.1 + \frac{0.4}{2} = 0.3$$

The power angle equation between V_t and V is

$$P = \frac{V_t V}{X_t} \sin \alpha$$

$$1.0 = \frac{1 \times 1}{0.3} \sin \alpha$$

or

$$\alpha = \sin^{-1} 0.3° = 17.458°$$

Taking terminal voltage of infinite bus as reference,

$$V_t = 1 \angle 17.458° = 0.954 + j0.3$$

The output current from the generator is

$$I = \frac{V_t - V}{jX_1}$$

$$= \frac{0.954 + j0.3 - 1}{j0.3} = 1 + j0.1535$$

The transient internal voltage E' is

$$E' = V_t + jIX_d' = 0.954 + j0.3 + j0.2(1 + j0.1535)$$
$$= 0.923 + j0.5 = 1.05 \angle 28.44°$$

The reactance between internal voltage and infinite bus is

$$X = 0.2 + X_1 = 0.2 + 0.3 = 0.5$$

The power angle equation relating the transient internal voltage E' and the infinite bus voltage V is

$$P_e = \frac{E'V}{X} \sin \delta = \frac{1.05 \times 1}{0.5} \sin \delta$$
$$= 2.1 \sin \delta$$

where δ is the machine rotor angle with respect to the infinite bus.

This power angle curve is plotted in Figure 9.10. Note that the mechanical input power P_m is constant and equals 1 pu, and the electrical power output intersects the sinusoidal power angle curve at the operating angle $\delta_0 = \sin^{-1}(1/2.1) = 28.44°$. This is the initial angular position of the generator rotor. The swing equation for the machine is

$$\frac{H}{180f} \frac{d^2\delta}{dt^2} = 1.0 - 2.1 \sin \delta$$

where

H = per unit inertia constant in MJ/MVA
f = electrical frequency of the system
δ is in elec. deg.

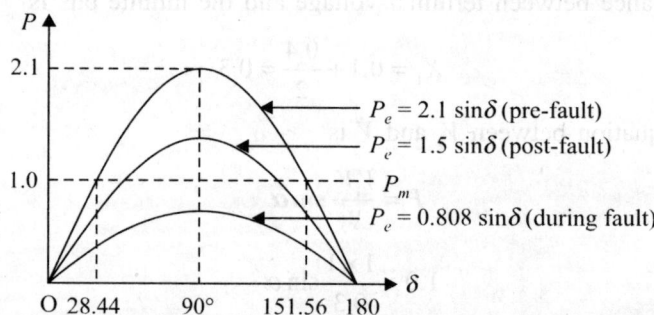

Figure 9.10   Power angle curves for Examples 9.3, 9.4 and 9.5.

EXAMPLE 9.4   The system of Example 9.3 is operating under the indicated conditions when a three-phase fault occurs at mid-point P of the line shown in Figure 9.8. Determine the power angle equation for the system with the fault on and the corresponding swing equation. Take $H = 5$ MJ/MVA.

Solution   The reactance diagram is shown in Figure 9.11(a) with the fault on the system at point P. The redrawn reactance diagram is shown in Figure 9.11(b). The reactance diagram after eliminating bus 3 by the conversion of Y into equivalent Δ is shown in Figure 9.11(c), where X is given by

$$X = \frac{0.3 \times 0.4 + 0.4 \times 0.2 + 0.2 \times 0.3}{0.2}$$
$$= \frac{0.12 + 0.08 + 0.06}{0.2} = 1.3$$

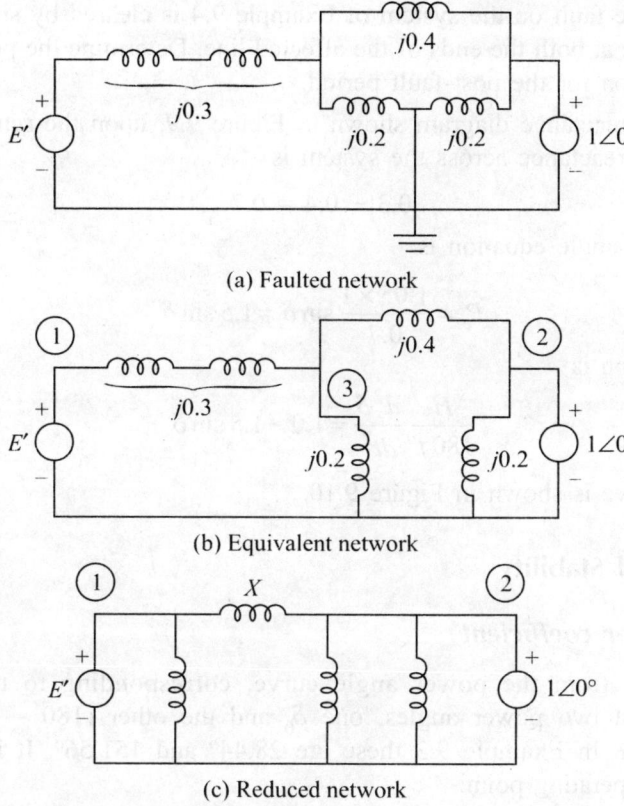

(a) Faulted network

(b) Equivalent network

(c) Reduced network

Figure 9.11 Reactance diagram for Example 9.4.

The power angle equation with the fault on is

$$P_e = \frac{E'V}{X} \sin \delta = \frac{1.05 \times 1}{1.3} \sin \delta = 0.808 \sin \delta$$

and the corresponding swing equation is

$$\frac{H}{180f} \frac{d^2\delta}{dt^2} = 1 - 0.808 \sin \delta$$

Substituting values of H and f, $\dfrac{d^2\delta}{dt^2} = 1800(1 - 0.808 \sin \delta)$ elec. deg./s^2

The power angle curve is shown in Figure 9.10.

Corresponding to the initial operating conditions,

$$P_m = P_{e0} = 1$$

and with fault the electrical power

$$P_e = 0.808 \sin \delta = 0.808 \sin 28.44° = 0.385$$

The accelerating power is

$$P_a = P_m - P_e = 1 - 0.385 = 0.615 \text{ pu}$$

EXAMPLE 9.5   The fault on the system of Example 9.4 is cleared by simultaneous opening of the circuit breakers at both the ends of the affected line. Determine the power angle equation and the swing equation for the post-fault period.

Solution   From the reactance diagram shown in Figure 9.9, upon the removal of the faulted line, the net transfer reactance across the system is

$$0.3 + 0.4 = 0.7$$

The post-fault power angle equation is

$$P_e = \frac{1.05 \times 1}{0.7} \sin \delta = 1.5 \sin \delta$$

and the swing equation is

$$\frac{H}{180f} \frac{d^2\delta}{dt^2} = 1.0 - 1.5 \sin \delta$$

The power angle curve is shown in Figure 9.10.

9.3.2   Small Signal Stability

Synchronizing power coefficient

It may be observed from the power angle curve, corresponding to the operating point $P_{e0} = P_m$, there exist two power angles, one δ_0 and the other $(180 - \delta_0)$ that satisfy the equilibrium condition. In Example 9.3 these are 28.44° and 151.56°. It is to be ascertained which is the stable operating point.

Consider a small temporary change in electrical output for a fixed mechanical input. Then,

$$\Delta P_e = \frac{dP_e}{d\delta}\bigg|_{\delta_0} \Delta\delta = P_{max} \cos \delta_0 \cdot \delta_\Delta \qquad \text{where } \delta\Delta = \Delta\delta \qquad (9.43)$$

The slope of the power angle curve about the operating point viz., change in electrical power output to change in power angle, is defined as synchronizing power coefficient, S_p and defined as

$$S_p = \frac{dP_e}{d\delta}\bigg|_{\delta=\delta_0} = P_{max} \cos \delta_0 \qquad (9.44)$$

Substituting in the swing equation, we have

$$\frac{2H}{\omega_s} \frac{d^2\delta_\Delta}{dt^2} + S_p \delta_\Delta = 0 \qquad \text{since } \Delta P_a = -\Delta P_e$$

$$\frac{d^2\delta_\Delta}{dt^2} + \frac{\omega_s S_p}{2H} \delta_\Delta = 0 \qquad (9.45)$$

This is a linear, second order differential equation, the solution of which depends upon the algebraic sign of S_p.

When S_p is positive, the solution $\delta_\Delta(t)$ corresponds to that of simple harmonic motion with undamped oscillations, as

$$\delta_\Delta \propto \sin \omega_n t \qquad (9.46)$$

where the angular frequency of the undamped sinusoidal oscillation is given by

$$\omega_n = \sqrt{\frac{\omega_s S_p}{2H}} \text{ elec. rad./s} \qquad (9.47a)$$

which corresponds to a frequency of oscillation

$$f_n = \frac{1}{2\pi} \sqrt{\frac{\omega_s S_p}{2H}} \text{ Hz} \qquad (9.47b)$$

and period of oscillation

$$T_n = \frac{1}{f_n} \text{ s} \qquad (9.47c)$$

When S_p is negative, the solution $\delta_\Delta(t)$ increases exponentially without limit.

Thus the operating point, $\delta_0 = 28.44°$ is a point of stable equilibrium, since S_p is positive and the rotor angle swing is bounded following a small disturbance and damping will restore the rotor angle to $\delta_0 = 28.44°$. On the other hand, $\delta_0 = 151.56°$ is not a stable operating point as S_p(slope of curve at this point) is negative.

The actual angular motion of the rotor is the superimposed motion of the swinging of rotor over the synchronous speed of the rotor.

EXAMPLE 9.6 A synchronous machine, having a maximum power transfer capability of 2.1 pu, is delivering 1 pu power, when it is subjected to a slight temporary electrical system disturbance. Determine the frequency and period of oscillation of the machine rotor if the disturbance is removed before the prime mover responds. $H = 5$ MJ/MVA.

Solution   The power angle equation is

$$P_e = P_{\max} \sin \delta = 2.1 \sin \delta$$

Initial operating power angle $\delta_0 = \sin^{-1}(1/2.1) = 28.44°$

The synchronizing power coefficient

$$S_p = P_{\max} \cos \delta_0 = 2.1 \cos 28.44° = 1.8466$$

The frequency of oscillation:

$$f_n = \frac{1}{2\pi} \sqrt{\frac{\omega_s S_p}{2H}} = \frac{1}{2\pi} \sqrt{\frac{2\pi \times 50 \times 1.8466}{2 \times 5}} = 1.2122 \text{ Hz}$$

The period of oscillation

$$T_n = \frac{1}{f_n} = 0.825 \text{ s}$$

In a large power system having many interconnected machines, the magnitude of the frequencies, which is superimposed upon the synchronous 50 Hz frequency, due to slow dynamics caused by load changes is of the order of 1–2 Hz. These oscillations are quickly damped out by the various damping influences caused by the prime mover, the system loads and the machine itself including its AVR.

9.4 Large Signal Stability or Transient Stability

9.4.1 Equal Area Criterion

Swing curves indicate whether a power system is stable or not after a disturbance. If the angle between any two machines tends to increase without limit, the system becomes unstable. If the angle tends to decrease with or without oscillations, then the system is stable. A simple graphical method of determining the stability of the system using power angle curves is called equal area criterion for stability.

Assumptions

1. Constant input power during the time interval is considered since mechanical devices take significant time to change input power.
2. Constant voltage behind transient reactance is assumed as flux change takes time.
3. Damping effect is neglected as it is going to improve the system stability in any case.

The equal area criterion is used for SMIB system. Although it can be adapted to two-machine system considering difference in angle between them as variable instead of absolute angles.

Suppose due to momentary disturbance and its removal, the rotor of the machine swings between the rotor angles δ_1 and δ_2 about the operating angle δ_0.

The swing equation is

$$\frac{2H}{\omega_s}\frac{d^2\delta}{dt^2} = P_m - P_e$$

Define the angular speed of rotor relative to synchronous speed by

$$\omega_r = \frac{d\delta}{dt} = \omega - \omega_s \tag{9.48}$$

Differentiating this and substituting in swing equation.

$$\frac{2H}{\omega_s}\frac{d\omega_r}{dt^2} = P_m - P_e$$

Multiplying both sides by $\omega_r = d\delta/dt$, we obtain

$$\frac{H}{\omega_s}2\omega_r\frac{d\omega_r}{dt} = (P_m - P_e)\omega_r$$

$$\frac{H}{\omega_s}\frac{d(\omega_r)^2}{dt} = (P_m - P_e)\frac{d\delta}{dt} \tag{9.49}$$

By integrating, we get

$$\frac{H}{\omega_s}(\omega_{r2}^2 - \omega_{r1}^2) = \int_{\delta_1}^{\delta_2}(P_m - P_e)d\delta \tag{9.50}$$

where

δ_1 and δ_2 = extreme angles about which the rotor swings

ω_{r1} = rotor speed at δ_1

ω_{r2} = rotor speed at δ_2

Since ω_r represents the departure of the rotor speed from synchronous speed and if the rotor speed is synchronous at δ_1 and δ_2, then correspondingly,

$$\omega_{r1} = \omega_{r2} = 0$$

Therefore, for stable system after the disturbance,

$$\int_{\delta_1}^{\delta_2} (P_m - P_e)d\delta = 0 \tag{9.51}$$

From the power angle curve shown in Figure 9.12, Eq. (9.51) can be rewritten as

$$\int_{\delta_1}^{\delta_0} (P_m - P_e)d\delta + \int_{\delta_0}^{\delta_2} (P_m - P_e)d\delta = 0$$

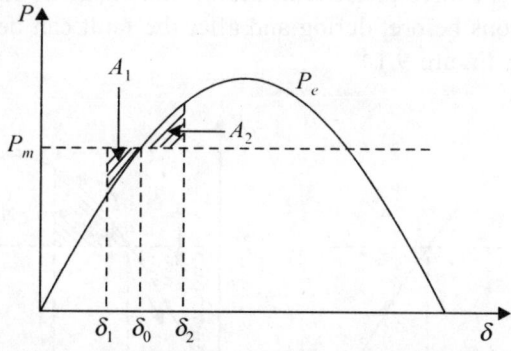

Figure 9.12   Power angle curve for equal area criterion.

It may be noted in the range $\delta_1 < \delta < \delta_0$, $P_e < P_{m,}$ say during the fault and in the range $\delta_0 < \delta < \delta_2$, $P_e > P_m$ (may be immediately after fault removal).

$$\int_{\delta 1}^{\delta 0} (P_m - P_e)d\delta = \int_{\delta 0}^{\delta 2} (P_e - P_m)d\delta \tag{9.52}$$

That is accelerating area equals decelerating area, which is the equal area criterion for stability. These areas are shown as A_1 and A_2 in Figure 9.12. Area A_1 is directly proportional to the increase in kinetic energy of the rotor while it is accelerating and Area A_2 is proportional to the decrease in kinetic energy of the rotor while it is decelerating.

9.4.2   Application of Equal Area Criterion

The equal area criterion can be applied to determine the following:

(i) Maximum sudden increase of generator input or load rejection.
(ii) Critical fault clearing angle and critical fault clearing time to maintain stability following a fault,

Only effect of three-phase fault on the system is studied.

Temporary fault at one end of the line

Consider the system shown in Figure 9.13. Initially, the system is operating at an angle δ_0 with P_m as the input power

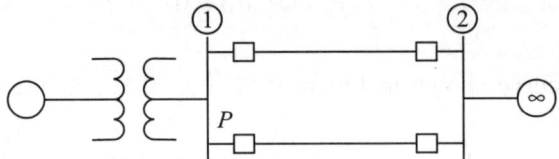

Figure 9.13 One-line diagram of a typical system showing the fault location P.

At the bus 1, a temporary three-phase fault occurs and clears by itself after a short period of time. The physical conditions before, during and after the fault can be understood by analysing the power angle curves in Figure 9.14.

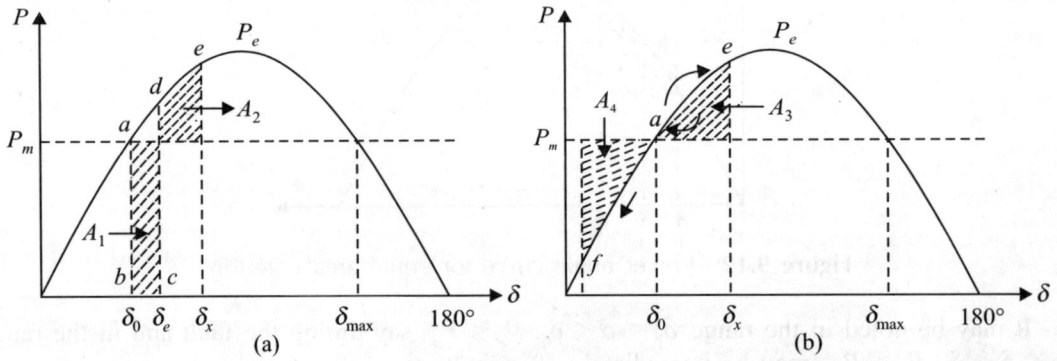

Figure 9.14 Power angle curves of generator before, during and after bus fault at P.

Originally, the generator is operating at synchronous speed with a rotor angle of δ_0 and the input mechanical power P_m equals the output electrical power P_e as shown at point a in Figure 9.14(a). When the fault occurs at $t = 0$, the electrical output suddenly becomes zero as shown at point b in Figure 9.14(a) while the input mechanical power remains the same. The difference in power must be accounted for by a rate of change of stored kinetic energy in the rotating masses. This can be accomplished only by an increase in speed, which results from the constant accelerating power P_m. If we denote the time to clear the fault by t_c, corresponding to δ_c, then for time $t < t_c$ the acceleration is given by

$$\frac{d^2\delta}{dt^2} = \frac{\omega_s}{2H} P_m \qquad (9.53)$$

While the fault is on, the speed-increase above synchronous speed is found by integrating this equation

$$\frac{d\delta}{dt} = \omega = \int_0^t \frac{\omega_s}{2H} P_m dt + \omega_s$$

or

$$\omega_r = \frac{\omega_s P_m}{2H} t \qquad (9.54)$$

A further integration with respect to time gives the rotor angle position

$$\delta = \frac{\omega_s P_m}{4H} t^2 + \delta_0 \qquad (9.55)$$

where δ_0 corresponds to the initial angle.

At the instant of fault clearing, the increase in rotor speed and the angle of separation between the generator and the infinite bus are

$$\omega_{rc} = \frac{\omega_s P_m}{2H} t_c \qquad (9.56)$$

$$\delta_c = \frac{\omega_s P_m}{4H} t_c^2 + \delta_0 \qquad (9.57)$$

When the fault clears at the angle δ_c, the electrical power output abruptly increases to a value corresponding to d on the power angle curve. At d the electrical output exceeds the mechanical power input and thus the accelerating power is negative or becomes decelerating (retarding). As a consequence the rotor slows down but rotor angle continues to increase as speed is still higher than synchronous speed when P_e goes from d to e in Figure 9.14(a). At e areas A_1 and A_2 are equal indicating that the energy gained due to acceleration is lost during deceleration and the rotor speed is again synchronous, although the rotor angle has advanced to δ_x. The accelerating power at e is still negative as $P_e > P_m$ and so the rotor cannot remain at synchronous speed but must continue to slow down. The relative speed now is negative and the rotor angle moves back from δ_x at e along the power angle curve of Figure 9.14(b) to point at a at which the rotor speed is less than synchronous speed. From a to f the mechanical power input exceeds the electrical power output and rotor increases speed again until it reaches synchronous speed at f. Point f is located so that areas A_3 and A_4 are equal. In the absence of damping, the rotor would continue to oscillate in the sequence $e - a - f, f - a - e$, etc. with synchronous speed occurring at e and f. In the real system, the damping influenced by the prime mover, loads and the machines damps out the oscillations to reach the steady state at a.

The equal area criterion states that whatever kinetic energy added to the rotor following a fault must be removed after the fault clearance to restore the rotor to synchronous speed.

Critical clearing angle and time: The shaded area A_1 is dependant upon the time taken to clear the fault. If there is delay in clearing, the angle δ_c is increased and the area A_1 increases. The equal area criterion requires that area A_2 also increase to restore the rotor to synchronous speed at a larger angle of δ_x. If the delay in clearing is prolonged so that the rotor angle

δ swings beyond the angle δ_{max}, which is $180 - \delta_0$, then positive accelerating power is again encountered while the rotor speed at that point on the power angle curve is already above synchronous speed. Under the influence of this positive accelerating power, the angle δ will increase without limit and instability results. Therefore, there is a critical angle for clearing the fault in order to satisfy the requirements of the equal area criterion for stability. This angle is called the *critical clearing angle* δ_{cr} as shown in Figure 9.15 and the corresponding critical time for removing the fault is called the *critical clearing time* t_{cr}.

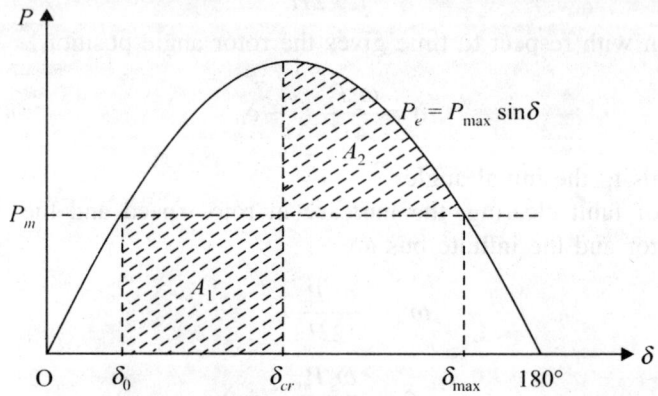

Figure 9.15 Power angle curve showing the critical clearing angle.

The areas A_1 and A_2 are given by

$$A_1 = \int_{\delta_0}^{\delta_{cr}} P_m d\delta = P_m(\delta_{cr} - \delta_0)$$

$$A_2 = \int_{\delta_{cr}}^{\delta_{max}} (P_{max} \sin\delta - P_m)d\delta$$

$$= P_{max}(\cos\delta_{cr} - \cos\delta_{max}) - P_m(\delta_{max} - \delta_{cr})$$

As per equal area criterion for stability,

$$A_1 = A_2$$

$$\cos\delta_{cr} = \frac{P_m}{P_{max}}(\delta_{max} - \delta_0) + \cos\delta_{max}$$

$$\delta_{max} = \pi - \delta_0 \tag{9.58}$$

and

$$P_m = P_{max} \sin\delta_0 \tag{9.59}$$

$\therefore$

$$\cos\delta_{cr} = (\pi - 2\delta_0) \sin\delta_0 - \cos\delta_0 \tag{9.60}$$

$$\delta_{cr} = \cos^{-1}[(\pi - 2\delta_0) \sin\delta_0 - \cos\delta_0] \tag{9.61}$$

From Eq. (9.57), we have

$$\delta_{cr} = \frac{\omega_s P_m}{4H} t_{cr}^2 + \delta_0 \qquad (9.62)$$

or

$$t_{cr} = \sqrt{\frac{4H(\delta_{cr} - \delta_0)}{\omega_s P_m}} \qquad (9.63)$$

EXAMPLE 9.7 A synchronous machine, having a maximum power transfer capability of 2.1 pu and H constant of 5 MJ/MVA, is delivering 1 pu power, when the system (Figure 9.13) is subjected to a temporary three-phase fault at bus 1. Calculate the critical clearing angle and the critical clearing time.

Solution The power angle equation is

$$P_e = P_{max} \sin \delta = 2.1 \sin \delta$$

Initial rotor angle is $\delta_0 = \sin^{-1}(1/2.1) = 28.44° = 0.496$ elec. rad.

$$\delta_{cr} = \cos^{-1}[(\pi - 2 \times 0.496) \sin 28.44° - \cos 28.44°]$$
$$= 81.697° = 1.426 \text{ elec. rad.}$$

$$t_{cr} = \sqrt{\frac{4 \times 5(1.426 - 0.496)}{2\pi \times 50}} = 0.2433 \text{ s}$$

This is equivalent to $0.2433/20 \times 10^{-3} = 12.16$ cycles based on frequency of 50 Hz (20 ms period).

Permanent fault at one end of the line with fault removed by line opening

Figure 9.16 shows a typical system with a three-phase solid fault occurring at P. This fault at the end of one of the lines, when cleared by opening of breakers at both ends will allow power to flow over the other parallel line.

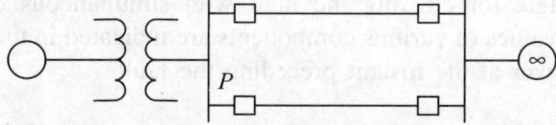

Figure 9.16 One-line diagram of a typical system indicating the location of fault P.

The power angle curves before, during and after the fault are shown in Figure 9.17. Before the fault, $P_{max0} \sin \delta$ is the power which can be transmitted; during the fault no power is transmitted and after the fault clearance at the instant when $\delta = \delta_{cr}$, $r_2 P_{max0} \sin \delta$ which is less than original power (as the transfer reactance is higher) is transmitted ($r_2 < 1$).

Applying equal area criterion

$$\int_{\delta_0}^{\delta_{cr}} P_m d\delta = \int_{\delta_{cr}}^{\delta_{max}} (r_2 P_{max0} \sin \delta - P_m) d\delta \qquad (9.64)$$

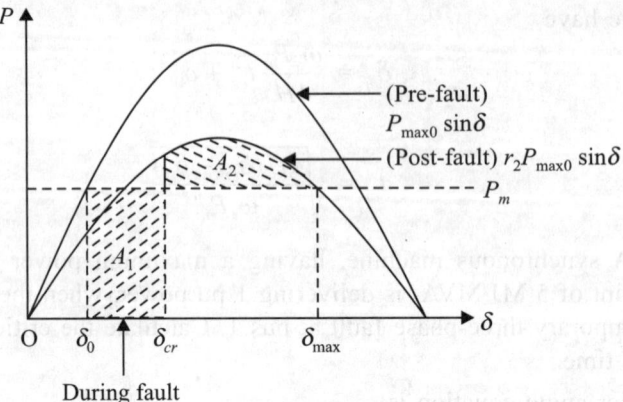

Figure 9.17 Equal area criterion applied to fault clearing for system shown in Figure 9.16.

$$P_m(\delta_{cr} - \delta_0) = r_2 P_{\max 0}[\cos \delta_{cr} - \cos \delta_{\max}] - P_m(\delta_{\max} - \delta_{cr})$$

$$P_{\max 0} \sin \delta_0 (\delta_{\max} - \delta_0) = r_2 P_{\max 0}(\cos \delta_{cr} - \cos \delta_{\max})$$

$$\cos \delta_{cr} = \frac{(\delta_{\max} - \delta_0) \sin \delta_0}{r_2} + \cos \delta_{\max}$$

where $\delta_{\max} = \pi - \sin^{-1} \dfrac{\sin \delta_0}{r_2}$ (since $r_2 P_{\max 0} \sin \delta = P_m = P_{\max 0} \sin_{\delta 0}$)
or

$$\delta = \sin^{-1} \frac{\sin \delta_0}{r_2} \tag{9.65}$$

$$t_{cr} = \sqrt{\frac{4H(\delta_{cr} - \delta_0)}{\omega_s P_m}} \tag{9.66}$$

EXAMPLE 9.8 A three-phase fault is applied at the point P as shown in Figure 9.18. Find the critical clearing angle for clearing the fault with simultaneous opening of the breakers 1 and 2. The reactance values of various components are indicated in the diagram. The generator is delivering 1.0 pu power at the instant preceding the fault.

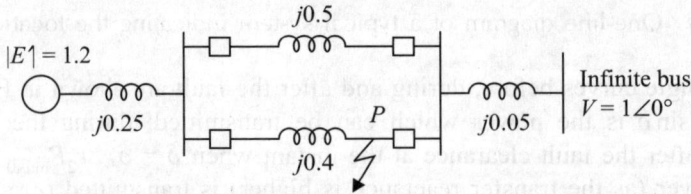

Figure 9.18 One-line diagram for Example 9.8.

Before fault

$$P_m = P_e = \frac{EV}{X_0} \sin \delta_0 = 1$$

where

$$X_{\text{pre-fault}} = 0.25 + \frac{0.5 \times 0.4}{0.5 + 0.4} + 0.05 = 0.522$$

$$\therefore \quad \frac{1.2 \times 1}{0.522} \sin \delta_0 = 1 \quad \text{or} \quad \sin \delta_0 = 0.435$$

$$\delta_0 = \sin^{-1}(0.435) = 25.79° = 0.45 \text{ rad.}$$

After fault clearance

$$X_{\text{post-fault}} = 0.25 + 0.5 + 0.05 = 0.8$$

$$r_2 = \frac{P_{\max 2}}{P_{\max 0}} = \frac{(1.2 \times 1)/0.8}{1.2 \times 1/0.522} = 0.653$$

$$\delta_{\max} = \pi - \sin^{-1} \frac{\sin \delta_0}{r_2}$$

$$= 180° - \sin^{-1} \frac{0.435}{0.653} = 180° - \sin^{-1}(0.67)$$

$$= 180° - 41.81° = 138.19° = 2.412 \text{ rad.}$$

$$\cos \delta_{cr} = \frac{(\delta_{\max} - \delta_0)\sin \delta_0}{r_2} + \cos \delta_{\max}$$

$$= (2.412 - 0.45) \times \frac{0.435}{0.653} - 0.766 = 0.541$$

$$\delta_{cr} = 57.25°$$

Fault at the middle of the line

When a three-phase fault occurs at some point on a double circuit line other than on the paralleling buses, there is some impedance between the paralleling buses and the fault. Therefore, some power is transmitted even when the fault is on the system. The power angle curve in Example 9.3 demonstrates this.

When power is transmitted during a fault, the equal area criterion is applied as shown in Figure 9.19. Before the fault, $P_{\max} \sin \delta$ is the power which can be transmitted; during the fault,

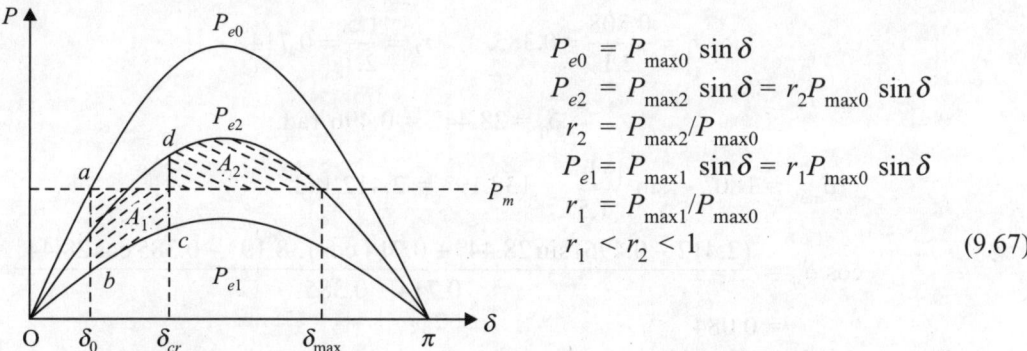

$$P_{e0} = P_{\max 0} \sin \delta$$
$$P_{e2} = P_{\max 2} \sin \delta = r_2 P_{\max 0} \sin \delta$$
$$r_2 = P_{\max 2}/P_{\max 0}$$
$$P_{e1} = P_{\max 1} \sin \delta = r_1 P_{\max 0} \sin \delta$$
$$r_1 = P_{\max 1}/P_{\max 0}$$
$$r_1 < r_2 < 1 \qquad (9.67)$$

Figure 9.19   Power angle curves of generator before, during and after the fault.

$r_1 P_{\max} \sin \delta$ is the power that can be transmitted and $r_2 P_{\max} \sin \delta$ is the power which can be transmitted after the fault is cleared by switching off the faulty line at the instant when $\delta = \delta_{cr}$, the critical clearing angle.

Note that $r_1 < r_2 < 1$. By evaluating the areas A_1 and A_2 using the equal area criterion,

$$\int_{\delta_0}^{\delta_{cr}} (P_m - P_{\max 1} \sin \delta) d\delta = \int_{\delta_{cr}}^{\delta_{\max}} (P_{\max 2} \sin \delta - P_m) d\delta$$

$$[P_m \delta + r_1 P_{\max 0} \cos \delta]_{\delta_0}^{\delta_{cr}} = [-r_2 P_{\max 0} \cos \delta - P_m \delta]_{\delta_{cr}}^{\delta_{\max}} \tag{9.68}$$

$$(\delta_{\max} - \delta_0) P_{\max 0} \sin \delta_0 + r_1 P_{\max 0} (\cos \delta_{cr} - \cos \delta_0) = -r_2 P_{\max 0} (\cos \delta_{\max} - \cos \delta_{cr})$$

since $P_m = P_{\max 0} \sin \delta_0$

$$\cos \delta_{cr} = \frac{(\delta_{\max} - \delta_0) \sin \delta_0 - r_1 \cos \delta_0 + r_2 \cos \delta_{\max}}{r_2 - r_1} \tag{9.69}$$

where

$$\delta_{\max} = \pi - \sin^{-1} \left(\frac{P_m}{P_{\max 2}} \right) = \pi - \sin^{-1} \left(\frac{\sin \delta_0}{r_2} \right) \tag{9.70}$$

$$t_{cr} = \sqrt{\frac{4H(\delta_{cr} - \delta_0)}{\omega_s P_m}} \tag{9.71}$$

EXAMPLE 9.9 Determine the critical clearing angle for the three-phase fault described in Examples 9.4 and 9.5 when the initial system configuration and pre-fault operating conditions are as described in Example 9.3.

Solution The power angle curves are shown in Figure 9.20 and the power angle equations are

Before the fault $\qquad P_{\max 0} \sin \delta \quad = 2.1 \sin \delta$

During the fault $\qquad r_1 P_{\max 0} \sin \delta = 0.808 \sin \delta$

After the fault removal $\qquad r_2 P_{\max 0} \sin \delta = 1.5 \sin \delta$

Hence,

$$r_1 = \frac{0.808}{2.1} = 0.385, \qquad r_2 = \frac{1.5}{2.1} = 0.714$$

$$\delta_0 = 28.44° = 0.496 \text{ rad.}$$

$$\delta_{\max} = 180° - \sin^{-1} \frac{1}{1.5} = 138.19° = 2.412 \text{ rad.}$$

$$\cos \delta_{cr} = \frac{(2.412 - 0.496) \sin 28.44° + 0.714 \cos 138.19° - 0.385 \cos 28.44°}{0.714 - 0.385}$$

$$= 0.084$$

$$\delta_{cr} = 85.181°$$

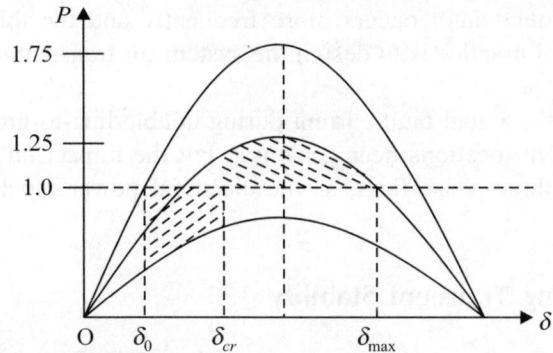

Figure 9.20 Power angle diagram for Example 9.9.

EXAMPLE 9.10 A three-phase generator delivers 1 pu power to an infinite bus through a transmission network when a fault occurs. The maximum power which can be transferred during pre-fault, during fault and post-fault conditions are 1.75 pu, 0.4 pu and 1.25 pu, respectively. Find the critical clearing angle.

Solution $P_m = 1$ pu, $P_{max0} = 1.75$, $P_{max1} = 0.4$, $P_{max2} = 1.25$

$$\delta_0 = \sin^{-1}\left(\frac{P_m}{P_{max\,0}}\right) = \sin^{-1}\left(\frac{1}{1.75}\right) = 34.85° = 0.608 \text{ rad.}$$

$$\delta_{max} = 180° - \sin^{-1}\left(\frac{P_m}{P_{max\,2}}\right) = 180° - \sin^{-1}\left(\frac{1}{1.25}\right) = 126.87° = 2.214 \text{ rad.}$$

$$\cos\delta_{cr} = \frac{1(2.214 - 0.608) - 0.4\cos 34.85° + 1.25\cos 126.87°}{1.25 - 0.4}$$

$$= 0.621$$

$$\delta_{cr} = 51.6°$$

9.4.3 Severity of Faults

Regardless of their location, short circuit faults not involving all three phases allow the transmission of some power, because they are represented by connecting some impedance rather than a short circuit between the fault point and the reference bus in the positive sequence impedance diagram. The larger the impedance shunted across the positive sequence network to represent the fault, the larger is the power transmitted during the fault. The amount of power transmitted during the fault affects the value of the accelerating area A_1 for any given clearing angle and correspondingly the decelerating area A_2 for stability. Smaller values of power transfer during fault results in larger A_1 and A_2 as the result of greater disturbances and the system tends towards instability. In the order of increasing severity, the various faults are:

- Single line-to-ground fault
- Double line-to-ground fault
- Line-to-line fault
- Three-phase fault

The single line-to-ground fault occurs more frequently and the three-phase fault is least frequent. But the universal practice is to design the system for transient stability for three phase faults at the worst locations.

It may be noted that the actual fault current during double line-to-ground fault is more than that for three-phase fault at locations near generator, but the impact on stability of the system is severe in the case of three-phase fault, as the electrical power blocked is more under this fault condition.

9.4.4 Factors Affecting Transient Stability

The type of fault and its location determines the severity of the fault and thereby the transient stability of a power system.

The two factors that affect the relative stability of a generating unit within a power system are:

(i) angular swing of the machine during and following fault clearance
(ii) critical clearing time.

The H constant and the transient reactance X_d' of the generating unit have a direct effect on these factors.

From the swing equation

$$\frac{d^2\delta}{dt^2} = \frac{\pi f}{H}(P_m - P_{max}\sin\delta) \qquad \text{where } P_{max} = \frac{EV}{X} \qquad (9.72)$$

It may be seen that lower inertia increases acceleration and also higher reactance.

Modern generators have better cooling arrangements because of which the MVA rating increases resulting in reduced H constant. Lower inertia requires lesser Short Circuit Ratio (SCR) and the reactance becomes larger. Thus the stability is affected both the ways. However, stability control techniques and transmission system designs have been evolved to increase overall system stability.

The control schemes are discussed as follows:

Fast excitation systems (static): When a fault occurs on a system the voltages at all buses are reduced. At generator terminals, the reduced voltages are sensed by the automatic voltage regulator (AVR), which acts within the excitation system to increase and restore generator terminal voltage thereby restricting the acceleration and the difference between the critical clearing angle and the initial operating angle, that is to reduce the initial rotor angle swing following the fault. The boost in the voltage applied to the field winding due to AVR action increases air gap flux that exerts a restraining torque on the rotor, which tends to slow down its motion. The effect is one-half to one and a half cycles gain in critical clearing time.

Turbine fast valve control: Modern electro hydraulic governing systems have the ability to close turbine valves to reduce unit acceleration during system faults near the unit. Immediately upon detecting that mechanical input is more than electrical output, control action initiates the valve closing which reduces the power input. A gain of one to two cycles in critical clearing time can be achieved.

Single pole operation and auto-reclosure: Reducing the transfer reactance of the system during fault conditions increases the power transfer decreasing the acceleration area, and thereby enhances the possibility of maintaining stability. Since single-phase faults occur more often than three-phase faults, protection schemes allowing independent circuit breaker pole operation can be used to clear the faulted phase while keeping the unfaulted phases intact. This increases the power transfer and can extend the critical clearing time by 2 to 5 cycles depending on type of fault.

Fast fault clearing times: Quicker breaker and high speed relaying can remove the fault faster. Faster the fault clearance from the system less is the acceleration area and better is the tendency of the system to restore to normal operating conditions.

Braking resistor at generator terminals: Braking resistors connected at the generator terminals act as resistive load reducing the accelerating power.

HVDC links: When operating in a synchronous mode that is in parallel with AC lines can modulate the power through the DC link very fast decreasing the accelerating power or damping the oscillations in the AC lines.

Short circuit current limiters: The form of series reactors can modify favourably the transfer impedance during fault conditions so that electrical power output can be increased enhancing the stability.

The system design which also improves steady-state stability:

- Minimum transformer reactance
- Series capacitor compensation
- Additional transmission lines

Series capacitor compensation and additional transmission lines: Reducing the reactance of transmission line is another way of raising P_{max}. Compensation for line reactance by series capacitors is often economical for increasing stability. Increasing the number of parallel lines is a common means of reducing reactance. When parallel lines are used, some power is transferred over the healthy lines even during a three-phase fault on one of the lines unless the fault occurs at a paralleling bus. Power transferred is subtracted from power input to obtain accelerating power. The increased power transfer during a fault thus reduces the accelerating power and the chance of instability.

9.5   Voltage Stability

Voltage instability implies an uncontrolled decrease in voltage monotonically, triggered by a disturbance, leading to voltage collapse and is primarily caused by the characteristics and dynamics of the connected load.

9.5.1   PV Curves

Consider a load met from infinite bus through transfer reactance X as shown in Figure 9.21. The MW load is increased in small steps while maintaining the power factor of the load constant.

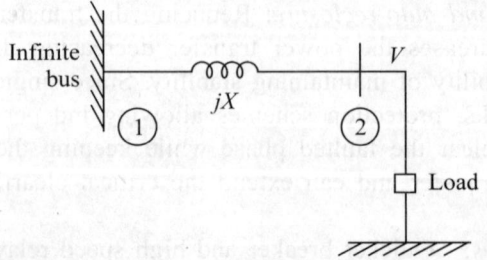

Figure 9.21   A two-bus system with radial load.

The slope dv/dp is monitored at each step and is checked to see if it is negative and has a value below a specified limit. The knee point where dv/dp is infinite gives the critical loading and any loading beyond this limit results in voltage instability and the load and voltage in fact decreases as shown in Figure 9.22. The distance between the operating point and the knee point gives the stability margin. The phasor diagram for system of Figure 9.21 is shown in Figure 9.23.

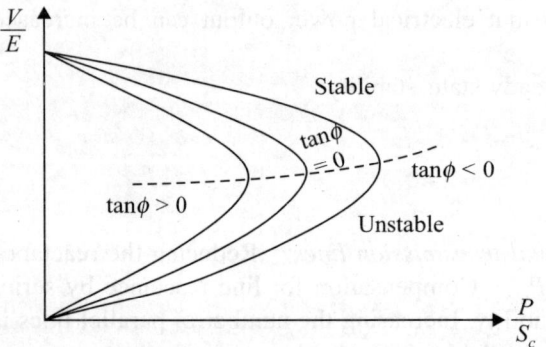

Figure 9.22   Normalised PV curves for different power factors.

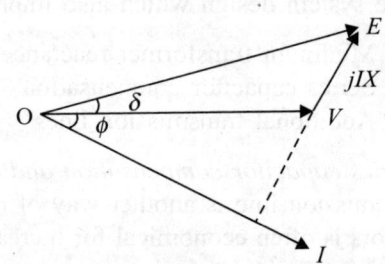

Figure 9.23   Phasor diagram for system of Figure 9.21.

$$V = \text{reference phasor}$$
$$E = E\angle\delta$$

In ΔOVE,

$$E^2 = V^2 + (IX)^2 - 2VIX \cos(90 + \phi) \tag{9.73}$$

$$= V^2 + \frac{S^2}{V^2}X^2 + 2VIX \sin\phi$$

$$= V^2 + \frac{P^2 + Q^2}{V^2}X^2 + 2QX$$

Multiplying both sides by V^2 and rearranging, we obtain

$$V^4 - (E^2 - 2QX)V^2 + X^2(P^2 + Q^2) = 0 \tag{9.74}$$

Dividing throughout by E^4 and using per unit values

$$v = \frac{V}{E} \tag{9.75}$$

$$p = \frac{P}{S_c} \tag{9.76}$$

$$q = \frac{Q}{S_c} \tag{9.77}$$

where

S_c is the short circuit level at load bus due to short circuit current I_{Sc} supplied by the infinite bus generator

$$= EI_{Sc} = E\frac{E}{X} = \frac{E^2}{X} \text{ where } E \text{ is the base voltage.} \tag{9.78}$$

$$v^4 - (1 - 2q)\, v^2 + p^2 + q^2 = 0 \tag{9.79}$$

Since $q = p \tan \phi$

$$v^4 - (1 - 2p \tan\phi)v^2 + (1 + \tan^2\phi)p^2 = 0 \tag{9.80}$$

To find maximum power, which corresponds to the critical power, the above equation is differentiated and dp/dv is equated to zero.

$$4v^3 - 2(1 - 2p \tan\phi)v + 2v^2 \tan\phi\frac{dp}{dv} + 2(1 + \tan^2\phi)p\,\frac{dp}{dv} = 0$$

$$2v_c^3 - (1 - 2p_{max} \tan \phi)\, v_c = 0$$

$$v_c[2v_c^2 - (1 - 2p_{max} \tan\phi)] = 0$$

$$v_c = \sqrt{\frac{1 - 2p_{max} \tan \phi}{2}} \tag{9.81}$$

v_c increases with improvement in power factor.

For unity power factor, $\cos\phi = 1$ and $\tan \phi = 0$.

$$v_c = \frac{1}{\sqrt{2}} = 0.707 \text{ pu} \tag{9.82}$$

Corresponding maximum power is given by Eq. (9.80)

$$p_{max}^2 = v_c^2 - v_c^4 = \frac{1}{2} - \frac{1}{4} = \frac{1}{4}$$

$$p_{max} = \frac{1}{2} = 0.5 \text{ pu} \tag{9.83}$$

As seen from Figure 9.22 the maximum power is higher when the power factor is leading. The critical voltage is also higher.

It may be remembered that if the power factor is improved with shunt capacitor compensation, any reduction in voltage is compounded because the reactive support decreases as square of the voltage and voltage collapse may occur.

9.5.2 Methods of Improving Voltage Stability

The following methods can improve the voltage stability and prevent voltage collapse:

- Installation of reactive power compensating devices at strategic locations
- Fast excitation system with AVR
- Transformer tap-changing
- Under-voltage load shedding

Short Answer Questions

1. **What is power system stability?**
   The stability of a power system is defined as the ability of the system to return to stable (synchronous) operation, when it is subjected to a disturbance.

2. **Define rotor angle stability.**
   Rotor angle stability is the ability of interconnected synchronous machines of a power system to remain in synchronism. The stability problem involves the study of the electro mechanical oscillations involving exchange of energy between network and generator mechanical system.

3. **Classify rotor angle stability.**
   The rotor angle stability phenomenon can be subdivided into two categories based on the severity/size of disturbances and the study period after disturbance.
   Large signal or transient stability
   Small signal or steady-state stability (static and dynamic)

4. **Define steady-state stability.**
   The steady-state stability of a power system is defined as the ability of the system to remain stable (without losing synchronism), when it is subjected to small disturbances such as gradual variation in rotor angle, voltage, etc.

5. **Define transient stability.**
   The transient stability of a power system is defined as the ability of the system to remain stable (without losing synchronism), when it is subjected to large disturbances such as occurrence of fault, sudden outage of a line or generator, sudden application or removal of large loads, etc.

6. **Define static stability.**
   Static stability is the inherent ability of the power system to return to a stable operating condition or remain in synchronism after a small disturbance without the aid of automatic control devices.

7. **Define dynamic stability.**
   Dynamic stability is the ability of the system to remain in synchronism when it is subjected to small disturbance such as oscillatory one (usually following transient state), with the help of automatic control devices.

8. Define mid-term and long-term stability.

   Mid-term stability comes into play when severe disturbances affect the system causing large frequency and voltage excursions. It involves both fast and slow dynamics and the study period is up to several minutes.

   Long-term stability is involved with slow dynamics and uniform frequency. The study period is up to tens of minutes.

9. Define voltage stability.

   Voltage instability implies an uncontrolled decrease in voltage monotonically, triggered by a disturbance, leading to voltage collapse and is primarily caused by the characteristics and dynamics of the connected load.

10. What is steady-state stability limit?

    The steady-state stability limit is the maximum power that can be transmitted to a system by a synchronous machine under steady-state without loss of synchronism. In steady-state the power transferred is always less than the steady-state stability limit.

11. What is transient stability limit?

    The transient stability limit is the maximum power that can be transmitted by a synchronous machine to a system during transient state without loss of synchronism. This is always less than the steady-state stability limit.

12. Indicate the units of inertia constants M and H and the interrelationship.

    The unit of M is MJ s/elect. rad.

    The unit of H is MJ/MVA or MW s/MVA or s.

    M and H are related by the equation

    $$M = \frac{HS}{\pi f}$$

    where
    S = MVA rating of machine
    f = frequency in Hz

13. What is the equivalent inertia constant, if two machines having inertia constants H_1 and H_2 (a) swing coherently and (b) swing noncoherently?

    When the inertia constants are on common base MVA, the equivalent inertia constant on the same MVA base is given by

    (a) $H_{eq} = H_1 + H_2$

    (b) $\dfrac{1}{H_{eq}} = \dfrac{1}{H_1} + \dfrac{1}{H_2}$

14. Define synchronizing coefficient. For what values of synchronizing coefficient the system remains stable?

    The synchronizing coefficient or stiffness of synchronous machine is given by

    $$P_{max} \sin \delta_0$$

    where $P_{max} = \dfrac{EV}{X}$

E = magnitude of internal emf of generator

V = magnitude of infinite bus voltage

X = transfer reactance between generator and infinite bus

δ_0 = operating power angle

The system is stable if the synchronizing coefficient is positive, that is when $0 \leq \delta_0 \leq \pi/2$.

15. Define power angle.

The power or torque angle is defined as the angular displacement of the rotor from synchronously rotating reference frame, which is the same between internal emf of generator and infinite bus.

16. Write the swing equation and explain the terms involved in it.

$$\frac{H}{\pi f} \frac{d^2\delta}{dt^2} = P_m - P_e$$

where

H = per unit inertia constant

δ = power angle in electrical radian

P_m = power supplied by the prime mover in pu

P_e = electrical power output in pu on the same base as H.

17. Define swing curve. What is its use?

The swing curve is the graph between the power angle and time.

It is usually plotted for transient state to study the nature of variation in power angle for a sudden large disturbance. From this the stability of the system can be determined.

18. Define critical clearing time and angle.

The critical clearing angle is the maximum allowable power angle before clearing the fault, without loss of synchronism. The time corresponding to this angle is called *critical clearing time*, which is also defined as the maximum time delay that can be allowed to clear a fault without loss of synchronism.

19. Give the expression for critical clearing time.

$$t_{cr} = \sqrt{\frac{4H(\delta_{cr} - \delta_0)}{\omega_s P_m}}$$

20. State equal area criterion.

The equal area criterion for stability states that the system is stable if the positive (acceleration) area under $P_e - \delta$ curve is equal to the negative (deceleration) area under $P_e - \delta$ curve for a finite change in δ.

21. List the methods of improving steady-state stability limit of a system.

See Section 9.3.1.

22. List the methods of improving transient stability limit of a power system.

- Increase of system voltage
- Fast excitation system (Static) with AVR
- Reduction in transfer reactance

- Turbine fast valve control
- Single pole operation and auto-reclosure
- Fast fault clearing time
- HVDC links
- Braking resistor at generator terminals

23. List the methods of improving voltage stability.

See Section 9.5.2.

Exercises

9.1 Find the critical clearing angle for clearing the fault shown in Figure E9.1 with simultaneous opening of the breakers 1 and 2. The reactance values of various components are indicated on the diagram. The generator is delivering 1.0 pu power at the instant preceding the fault. The fault occurs at point P as shown in the Figure.

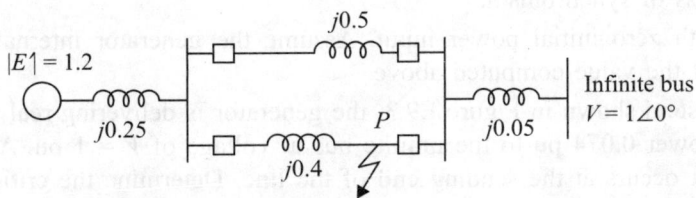

Figure E9.1   One-line diagram for Problem 9.1.

9.2 The generator shown in Figure E9.2 is delivering power to infinite bus. Take $V_t = 1.1$ pu. Find the maximum power that can be transferred when the system is healthy.

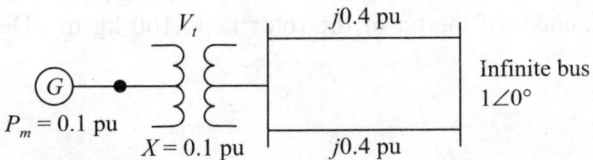

Figure E9.2   One-line diagram for Problem 9.2.

9.3 A 50 Hz generator is delivering 50% of the power that it is capable of delivering through a transmission line to an infinite bus. A fault occurs that increases the reactance between the generator and the infinite bus to 500% of the value before the fault. When the fault is isolated, the maximum power that can be transferred is 75% of the original maximum value. Determine the critical clearing angle for the condition described.

9.4 Consider a synchronous machine characterized by the parameter $X_d = 1.0$, $X_d' = 0.3$ pu. The machine is connected directly to an infinite bus of voltage 1.0 pu. The generator is delivering a real power of 0.5 pu at 0.8 power factor lagging. Determine the voltage behind transient reactance and the transient power angle equation.

[**Ans.** $1.1226\angle 7.68°$, $3.7419 \sin \delta$]

9.5 A 50 Hz synchronous generator having inertia constant $H = 9.94$ MJ/MVA and a transient reactance $X'_d = 0.3$ pu is connected to an infinite bus through a purely reactive circuit as shown in Figure E9.3. The generator is delivering real power of 0.6 pu, 0.8 power factor lagging to the infinite bus at a voltage of $V = 1$ pu. Find out the frequency of oscillation of the generator, if the breakers open and then close quickly.

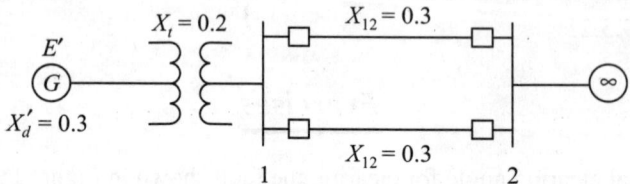

Figure E9.3 Circuit for Problem 9.5.

9.6 For the machine of Problem 9.5 determine the maximum power input that can be applied without loss of synchronism. [**Ans.** 1.084 pu]

Repeat with zero initial power input. Assume the generator internal voltage remains constant at the value computed above. [**Ans.** 1.505 pu]

9.7 For the system shown in Figure E9.3, the generator is delivering real power 0.8 pu and reactive power 0.074 pu to the infinite bus at voltage of $V = 1$ pu. A temporary three-phase fault occurs at the sending end of the line. Determine the critical clearing angle and the critical fault clearing time. [**Ans.** 84.775°, 0.285 s]

9.8 Repeat Problem 9.7 if the fault occurs at the middle of one of the lines and the fault is cleared by isolating the faulted line. Determine the critical clearing angle.
[**Ans.** 98.834°]

9.9 A four pole 50 Hz synchronous generator has a rating of 200 MVA, 0.8 power factor lagging. The moment of inertia of the rotor is 45,100 kg m². Determine M and H.
[**Ans.** 7.1, 2.8]

CHAPTER 10

Numerical Methods for Solving Swing Equation

Direct plotting of rotor angles of generators with respect to time can demonstrate how the machines swing with each other and can be computed from swing equations. Numerical methods are convenient and amenable to using computers for solving the swing equations to obtain the rotor angles.

10.1 Solution of Swing Equation

The following numerical methods illustrate the procedures adopted for solving the swing equations to obtain the rotor angles:

- Euler's method
- Modified Euler's method
- Runge–Kutta method

10.1.1 Euler's Method

Consider the following equation:

$$\frac{dx}{dt} = f(x, t) \tag{10.1}$$

where t and x are the independent and dependent variables, respectively. If this equation can be integrated exactly, then the solution is of the form

$$x = \phi(t) \tag{10.2}$$

This equation may represent a smooth curve as shown in Figure 10.1.

If the function is expanded in Taylor's series about the operating point x_0.

$$x = x_0 + \dot{x}_0 \Delta t + \ddot{x}_0 \frac{(\Delta t)^2}{2} + \cdots \tag{10.3}$$

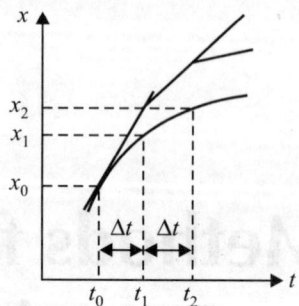

Figure 10.1   Graph of approximate solution by Euler's method.

Neglecting higher order terms beyond first derivative, we get

$$x_1 = x_0 + \dot{x}_0 \Delta t \tag{10.4}$$

where,

$$\dot{x}_0 = \frac{dx}{dt}\bigg|_{x_0}$$

Repeating the process, we have

$$x_2 = x_1 + \dot{x}_1 \Delta t$$

$$\vdots \qquad \vdots \qquad \vdots$$

$$x_{n+1} = x_n + \dot{x}_n \Delta t$$

We can apply this principle for solving swing equation to get the rotor angle. As our interest is only on change in rotor angle, which depends on the relative rotor speed with reference to the synchronous speed we consider only the relative speed for computation purpose. The second order swing equation can be split into two first order differential equations with δ and ω_r being state variables.

$$\frac{d^2\delta}{dt^2} = \frac{\pi f}{H}(P_m - P_e)$$

$$\frac{d\omega_r}{dt} = \frac{\pi f}{H}(P_m - P_e) \tag{10.5}$$

since

$$\frac{d\delta}{dt} = \omega_r = \omega - \omega_s \tag{10.6}$$

At the time of fault

$$\delta_0 = \sin^{-1}\left(\frac{P_m}{P_{e0\,max}}\right)$$

$$\omega_{r0} = 0 \text{ (since } \omega_0 = \omega_s)$$

$$\omega_{r0} = \frac{\pi f}{H}(P_m - P_{e0}) \text{ and } P_{e0} < P_m \text{ due to the occurrence of fault.}$$

Rotor angle δ depends on rotor speed change ω_r, which in turn depends on δ through $\dot{\omega}_r$ and P_e and so on cyclically.

Applying Euler's method, we obtain

$$\delta_1 = \delta_0 + \left.\frac{d\delta}{dt}\right|_0 \Delta t = \delta_0 + \omega_{r0}\Delta t \tag{10.7}$$

and

$$\omega_{r1} = \omega_{r0} + \left.\frac{d\omega_r}{dt}\right|_0 \Delta t = 0 + \dot{\omega}_{r0}\Delta t \tag{10.8}$$

Euler's method suffers from the following shortcomings as it uses derivative at the beginning only:

   (i) The error increases as we take increasing values of t as can be seen in Figure 10.1.

   (ii) If the step size Δt is large, the method may give erroneous results.

10.1.2 Modified Euler's Method

The drawbacks of Euler's method can be partially overcome by the use of modified Euler's method. Average of the derivatives at the beginning and the end is used to find the value. The steps in this method are the predictor and corrector steps.

$$\text{Predictor step } x_1^P = x_0 + \dot{x}_0 \Delta t \tag{10.9}$$

$$\text{Corrector step } x_1^C = x_0 + \frac{1}{2}[\dot{x}_0 + \dot{x}_1^P]\Delta t \tag{10.10}$$

These two steps can be repeated until two successive estimates are within specified tolerance. From swing equations,

$$\delta_0 = \sin^{-1}\left(\frac{P_m}{P_{e0\,max}}\right)$$

The derivatives at the beginning of the step are:

$$\left.\frac{d\delta}{dt}\right|_0 = \omega_{r0} = 0 \quad (\text{since } \omega_0 = \omega_s)$$

$$\left.\frac{d\omega_r}{dt}\right|_{0+} = \dot{\omega}_{r0} = \frac{\pi f}{H}(P_m - P_{e0})$$

At the end of first step

$$\delta_1^P = \delta_0 + \dot{\delta}_0 \Delta t = \delta_0 + \omega_{r0}\Delta t \tag{10.11}$$

$$\omega_{r1}^P = \omega_{r0} + \dot{\omega}_{r0}\Delta t \tag{10.12}$$

Using the predicted values of δ_1^P and ω_{r1}^P, the derivatives at the end of 1st interval are determined by

$$\left.\frac{d\delta_1^P}{dt}\right|_{\omega_{r1}^P} = \omega_{r1}^P$$

$$\left.\frac{d\omega_{r1}^P}{dt}\right|_{\delta_1^P} = \dot{\omega}_{r1}^P = \frac{\pi f}{H}(P_m - P_{e1})$$

The average value of the two derivatives is then used to find the corrected value.

$$\delta_1^C = \delta_0 + \frac{1}{2}[\dot{\delta}_0 + \dot{\delta}_1^P]\Delta t = \delta_0 + \frac{1}{2}[\omega_{r0} + \omega_{r1}^P]\Delta t \tag{10.13}$$

$$\omega_{r1}^C = \omega_{r0} + \frac{1}{2}[\dot{\omega}_{r0} + \dot{\omega}_{r1}^P]\Delta t \tag{10.14}$$

The flow chart for this method is given in Figure 10.2.

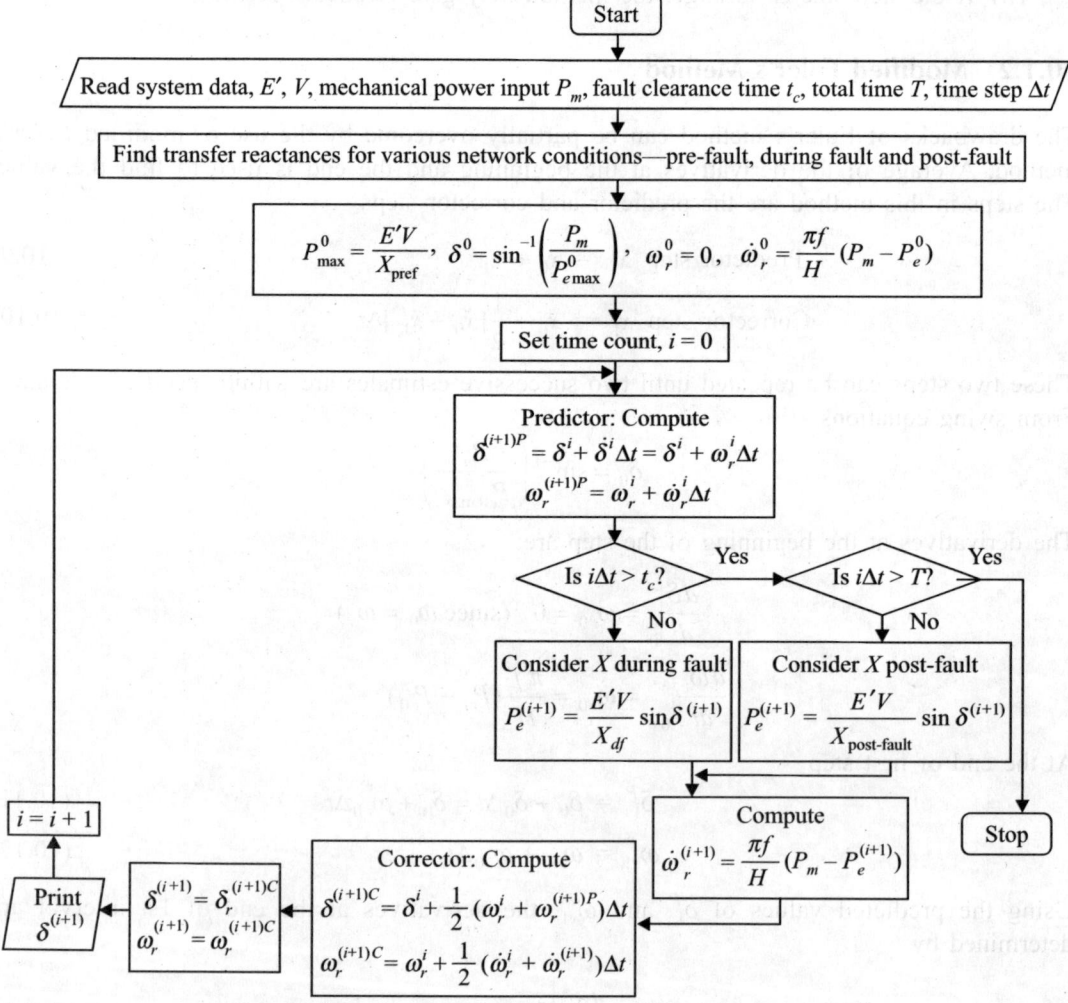

Figure 10.2   Flow chart for modified Euler's method.

10.1.3 Runge–Kutta Method

This method also approximates the Taylor's series.

Second order Runge–Kutta method

This method is similar to Modified Euler's method and gives more accurate results than Euler's method.

$$\frac{dx}{dt} = f(x,\ t) \tag{10.15}$$

$$x_1 = x_0 + \Delta x = x_0 + \frac{m_1 + m_2}{2} \tag{10.16}$$

where,

$$m_1 = \dot{x}_0 \Delta t = f(x_0, t_0)\Delta t \tag{10.17}$$

$$m_2 = f(x_0 + m_1,\ t_0 + \Delta t)\Delta t \tag{10.18}$$

In general,

$$x_{n+1} = x_n + \frac{m_1 + m_2}{2} \tag{10.19}$$

where

$$m_1 = f(x_n\ t_n)\Delta t \tag{10.20}$$
$$m_2 = f(x_n + m_1,\ t_n + \Delta t)\Delta t \tag{10.21}$$

From Swing equations, we have

$$\delta_0 = \sin^{-1}\left(\frac{P_m}{P_{e0\,\max}}\right)$$

$$\omega_{r0} = 0 \quad (\text{since } \omega_0 = \omega_s)$$

$$\delta_1 = \delta_0 + \frac{k_1 + k_2}{2} \tag{10.22}$$

$$\omega_{r1} = \omega_{r0} + \frac{l_1 + l_2}{2} \tag{10.23}$$

where

$$k_1 = \left.\frac{d\delta}{dt}\right|_{\omega_{r0},t_0} \Delta t = \omega_{r0}\Delta t \tag{10.24}$$

$$l_1 = \left.\frac{d\omega_r}{dt}\right|_{\delta_0,t_0} \Delta t = \frac{\pi f}{H}(P_m - P_{e0})\Delta t \tag{10.25}$$

$$k_2 = \left.\frac{d\delta}{dt}\right|_{\ell_1,t_0+\Delta t} \Delta t = l_1 \Delta t \tag{10.26}$$

and

$$l_2 = \left.\frac{d\omega_r}{dt}\right|_{\delta_0+k_1,t_0+\Delta t} \Delta t = \frac{\pi f}{H}(P_m - P_{e(\delta_0+k_1)})\Delta t \tag{10.27}$$

The flow chart for this method is given in Figure 10.3.

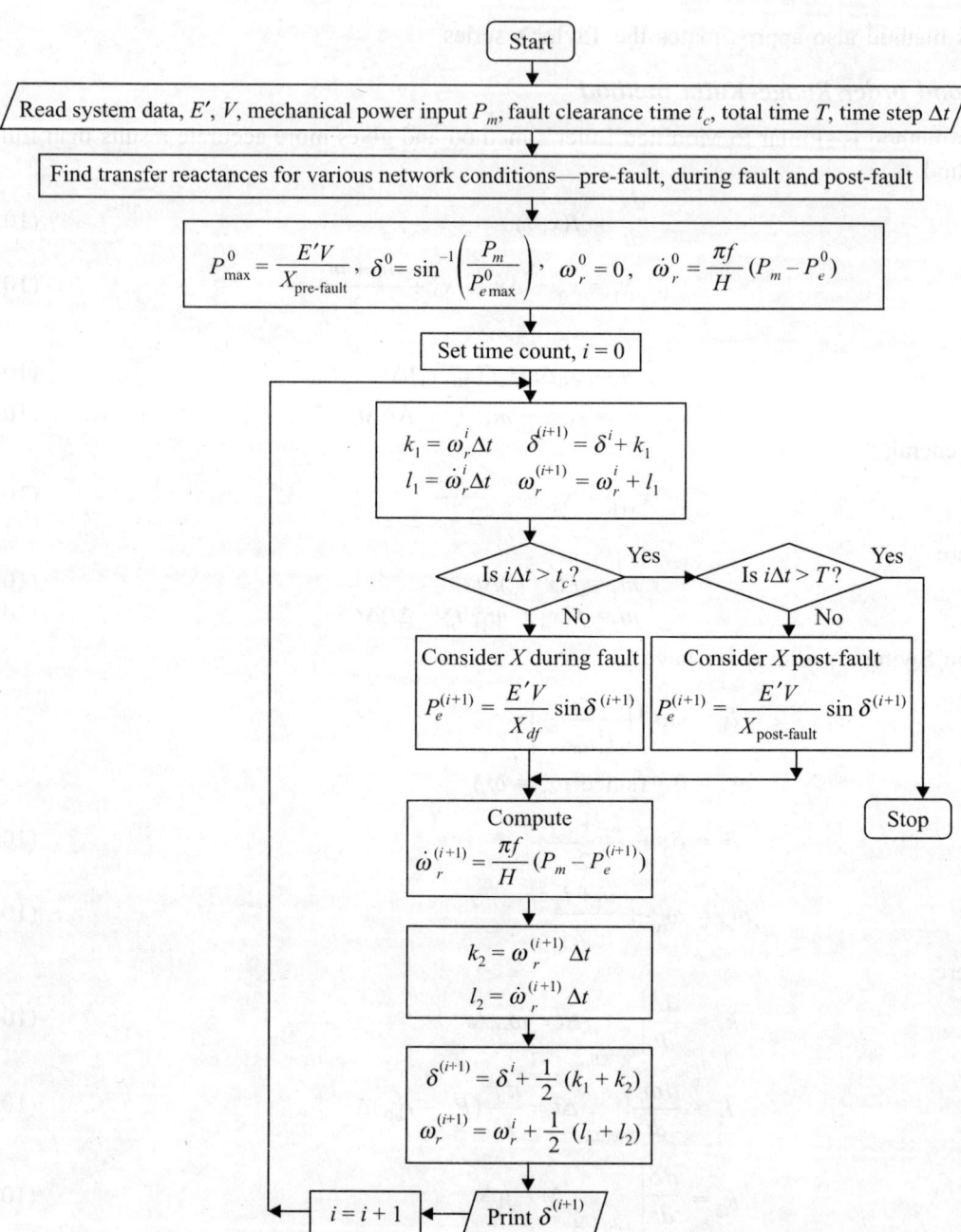

Figure 10.3   Flow chart for Runge–Kutta second order method.

Fourth order Runge–Kutta method

This method gives more accurate results than Euler's method, modified Euler's method and second order Runge–Kutta method.

$$x_{n+1} = x_n + \frac{1}{6}(m_1 + 2m_2 + 2m_3 + m_4) \tag{10.28}$$

where

$m_1 = $ (Slope at the beginning of time step)$\Delta t = f(x_n, t_n)\Delta t$ (10.29)

$m_2 = $ (First approximation to derivative at mid-step) Δt

$$= f\left(x_n + \frac{k_1}{2}, t_n + \frac{\Delta t}{2}\right)\Delta t \tag{10.30}$$

$m_3 = $ (Second approximation to derivative at mid-step) Δt

$$= f\left(x_n + \frac{k_2}{2}, t_n + \frac{\Delta t}{2}\right)\Delta t \tag{10.31}$$

$m_4 = $ (Derivative at the end of time step) Δt
$$= f(x_n + k_3, t_n + \Delta t)\,\Delta t \tag{10.32}$$

$$\Delta x = \frac{1}{6}(m_1 + 2m_2 + 2m_3 + m_4) \tag{10.33}$$

which is weighted average of estimates based on derivatives at the beginning, mid-point and end of time step.

From Swing equations, we obtain

$$\delta_0 = \sin^{-1}\left(\frac{P_m}{P_{e0\,max}}\right)$$

$$\omega_{r0} = 0 \qquad (\text{since } \omega_0 = \omega_s)$$

$$\delta_1 = \delta_0 + \frac{1}{6}(k_1 + 2k_2 + 2k_3 + k_4) \tag{10.34}$$

$$\omega_{r1} = \omega_{r0} + \frac{1}{6}(l_1 + 2l_2 + 2l_3 + l_4) \tag{10.35}$$

where

$$k_1 = \left.\frac{d\delta}{dt}\right|_{\omega_{r0},t_0} \Delta t = \omega_{r0}\Delta t \tag{10.36}$$

$$l_1 = \left.\frac{d\omega_r}{dt}\right|_{\delta_0,t_0} \Delta t = \frac{\pi f}{H}(P_m - P_{e0})\,\Delta t \tag{10.37}$$

$$k_2 = \left.\frac{d\delta}{dt}\right|_{\omega_{r0}+\frac{l_1}{2},t_0+\frac{\Delta t}{2}} \Delta t = \frac{l_1}{2}\Delta t \tag{10.38}$$

$$l_2 = \frac{d\omega_r}{dt}\bigg|_{\delta_0 + \frac{k_1}{2}, t_0 + \frac{\Delta t}{2}} \Delta t = \frac{\pi f}{H}\left[P_m - P_{e\left(\delta_0 + \frac{k_1}{2}\right)}\right]\Delta t \qquad (10.39)$$

$$k_3 = \frac{d\delta}{dt}\bigg|_{\omega_{r0} + \frac{l_2}{2}, t_0 + \frac{\Delta t}{2}} \Delta t = \frac{l_2}{2}\Delta t \qquad (10.40)$$

$$l_3 = \frac{d\omega_r}{dt}\bigg|_{\delta_0 + \frac{k_2}{2}, t_0 + \frac{\Delta t}{2}} \Delta t = \frac{\pi f}{H}\left[P_m - P_{e\left(\delta_0 + \frac{k_2}{2}\right)}\right]\Delta t \qquad (10.41)$$

$$k_4 = \frac{d\delta}{dt}\bigg|_{\omega_{r0} + l_3, t_0 + \Delta t} \Delta t = l_3 \Delta t \qquad (10.42)$$

$$l_4 = \frac{d\omega_r}{dt}\bigg|_{\delta_0 + k_3, t_0 + \Delta t} \Delta t = \frac{\pi f}{H}[P_m - P_{e(\delta_0 + k_3)}]\Delta t \qquad (10.43)$$

EXAMPLE 10.1 A three-phase generator delivers 1 pu power to an infinite bus through a transmission network when a three-phase fault occurs at the generator bus. The maximum power that can be transferred before fault is 4.325 pu. Find the rotor angle and speed after a time step of 0.02 s by Euler, modified Euler and Runge–Kutta methods. Take $H = 8$ MJ/MVA.

Solution During the fault conditions, the mechanical input power P_m remains constant at 1 pu and electrical power output P_e is zero.

Initial conditions:

$$P_m = P_{e0\max} \sin \delta_0$$

$$\sin \delta_0 = \frac{P_m}{P_{e0\max}} = \frac{1}{4.325}$$

$$\delta_0 = 13.365° = 0.2333 \text{ rad.}$$

$$\omega_s = 2\pi f = 314.16 \text{ rad./s and } \omega_{r0} = \omega_0 - \omega_s = 0$$

$$\dot{\omega}_{r0} = \frac{\pi f}{H}(P_m - P_{e0}) = \frac{\pi \times 50}{8}(1 - 0) = 19.63$$

Euler's method:

$$\delta_1 = \delta_0 + \frac{d\delta}{dt}\bigg|_0 \Delta t = \delta_0 + \omega_{r0}\Delta t = 0.2333 \text{ rad.} = 13.365°$$

$$\omega_{r1} = \omega_{r0} + \frac{d\omega_r}{dt}\bigg|_{0+} \Delta t = 0 + \dot{\omega}_{r0}\Delta t = 19.63 \times 0.02 = 0.393$$

$$\omega_1 = \omega_s + \omega_{r1} = 314.16 + 0.393 = 314.553 \text{ rad./s}$$

Modified Euler's method:

Predictor step

$$\delta_1^P = \delta_0 + \dot{\delta}_0 \Delta t = \delta_0 + \omega_{r0}\Delta t = 0.2333 + 0 = 0.2333 \text{ rad.}$$

$$\omega_{r1}^P = \omega_{r0} + \dot{\omega}_{r0}\Delta t = 19.63 \times 0.02 = 0.393 \text{ rad./s}$$

Corrector step

$$\delta_1^C = \delta_0 + \frac{1}{2}[\dot{\delta}_0 + \dot{\delta}_1^P]\Delta t = \delta_0 + \frac{1}{2}[\omega_{r0} + \omega_{r1}^P]\Delta t$$

$$= 0.2333 + \frac{1}{2}(0 + 0.1963) \times 0.02 = 0.2353 \text{ rad.} = 13.482°$$

$$\dot{\omega}_{r1}^P = \frac{d\omega_r}{dt}\bigg|_{\delta_1^P} = \frac{\pi f}{H}(P_m - P_{e1}) = \frac{\pi \times 50}{8}(1 - 0) = 19.63$$

$$\omega_{r1}^C = \omega_{r0} + \frac{1}{2}[\dot{\omega}_{r0} + \dot{\omega}_{r1}^P]\Delta t = 0.393 \text{ rad./s}$$

$$\omega_1^C = \omega_s + \omega_{r1}^C = 314.16 + 0.393 = 314.553 \text{ rad./s}$$

Runge–Kutta method:

Second order Runge–Kutta method:

First estimate:

$$k_1 = \frac{d\delta}{dt}\bigg|_0 \Delta t = \omega_{r0}\Delta t = 0$$

$$l_1 = \frac{d\omega_r}{dt}\bigg|_{0+} \Delta t = \frac{\pi f}{H}(P_m - P_{e0})\Delta t = \frac{\pi \times 50}{8}(1 - 0) \times 0.02 = 0.393$$

Second estimate:

$$k_2 = l_1\Delta t = 0.393 \times 0.02 = 0.00786$$

$$l_2 = \frac{d\omega_r}{dt}\bigg|_{\delta_0+k_1} \Delta t = \frac{\pi f}{H}[P_m - P_{e(\delta_0+k_1)}] \times 0.02 = 0.393$$

$P_{e(\delta_0+k_1)}$ is the machine power obtained from power flow using $\delta_1 = \delta_0 + k_1$.

Since the fault is on the terminal of the generator $P_{e(\delta_0+k_1)} = 0$.

$$\delta_1 = 0.2333 + \frac{0 + 0.00786}{2} = 0.2372 \text{ rad.} = 13.588°$$

$$\omega_{r1} = \omega_{r0} + \frac{l_1 + l_2}{2} = 0.393$$

$$\omega_1 = \omega_s + \omega_{r1} = 314.16 + 0.393 = 314.553 \text{ rad./s}$$

Fourth order Runge–Kutta method:

First estimate is the same as calculated in second order method.

Second estimate:

$$k_2 = \frac{l_1}{2}\Delta t = \frac{0.393}{2} \times 0.02 = 0.00393$$

$$l_2 = \frac{\pi f}{H}[P_m - P_{e(\delta_0+k_1/2)}] \times 0.02 = 0.393$$

Third estimate:

$$k_3 = \frac{l_2}{2} \Delta t = \frac{0.393}{2} \times 0.02 = 0.00393$$

$$l_3 = \frac{\pi f}{H} [P_m - P_{e(\delta_0 + k_2/2)}] \Delta t = 0.393$$

Fourth estimate:

$$k_4 = l_3 \Delta t = 0.393 \times 0.02 = 0.00786$$

$$l_4 = \frac{\pi f}{H} [P_m - P_{e(\delta_0 + k_3)}] \times 0.02 = 0.393$$

Final estimates:

$$\delta_1 = \delta_0 + \frac{1}{6}(k_1 + 2k_2 + 2k_3 + k_4)$$

$$= 0.2333 + \frac{1}{6}(0 + 0.00786 + 0.00786 + 0.00786)$$

$$= 0.2372 \text{ rad.} = 13.59°$$

$$\omega_{r1} = \omega_{r0} + \frac{1}{6}(l_1 + 2l_2 + 2l_3 + l_4)$$

$$= 0 + \frac{1}{6}(0.393 + 0.786 + 0.786 + 0.393) = 0.393 \text{ rad./s}$$

$$\omega_1 = \omega_s + \omega_{r1} = 314.16 + 0.393 = 314.553 \text{ rad./s}$$

EXAMPLE 10.2 A three-phase generator is delivering 0.8 pu power when a fault occurs in the network reducing the maximum power transfer capability from 1.932 pu to 0.742 pu. Obtain the numerical solution of the swing equation after a time step of 0.05 s, using Euler, modified Euler and Runge–Kutta methods. Take $H = 5$ MJ/MVA.

Solution

$$P_m = P_{e0} = P_{e0max} \sin \delta_0$$

$$\sin \delta_0 = \frac{P_m}{P_{e0\,max}} = \frac{0.8}{1.932} = 0.414$$

$$\delta_0 = 0.427 \text{ rad.} = 24.456°$$

$$P_{e0+} = 0.742 \sin \delta$$

where suffix '0+' indicates immediately after fault.

Euler's method:

$$\delta_1 = \delta_0 + \frac{d\delta}{dt}\bigg|_0 \Delta t = \delta_0 + \omega_{r0} \Delta t$$

$$= 0.427 \text{ rad.} = 24.456°$$

$$\omega_{r1} = \omega_{r0} + \frac{d\omega_r}{dt}\bigg|_0 \Delta t = 0 + \dot{\omega}_{r0}\Delta t$$

$$= \frac{\pi f}{H}(P_m - P_{e0+})\Delta t = \frac{\pi f}{H}(0.8 - 0.742\sin 0.427) = 15.479$$

Modified Euler's method:
Predictor step

$$\delta_1^P = \delta_0 + \dot{\delta}_0\Delta t = \delta_0 + \omega_{r0}\Delta t = 0.427 + 0 = 0.427 \text{ rad.}$$

$$\omega_{r1}^P = \omega_{r0} + \dot{\omega}_{r0}\Delta t = 15.479 \times 0.05 = 0.774 \text{ rad./s}$$

Corrector step

$$\delta_1^C = \delta_0 + \frac{1}{2}[\dot{\delta}_0 + \dot{\delta}_1^P]\Delta t = \delta_0 + \frac{1}{2}[\omega_{r0} + \omega_{r1}^P]\Delta t$$

$$= 0.427 + \frac{1}{2}(0 + 0.387) \times 0.05 = 0.4367 \text{ rad.} = 25.021°$$

$$\dot{\omega}_{r1}^P = \frac{d\omega_r}{dt}\bigg|_{\delta_1^P} = \frac{\pi f}{H}(P_m - P_{e1}) = \frac{\pi f}{H}(0.8 - 0.742\sin 0.427) = 15.479$$

$$\omega_{r1}^C = \omega_{r0} + \frac{1}{2}[\dot{\omega}_{r0} + \dot{\omega}_{r1}^P]\Delta t = 0.387 \text{ rad./s}$$

Runge–Kutta method:
Second order Runge–Kutta Method:
First estimate:

$$k_1 = \frac{d\delta}{dt}\bigg|_0 \Delta t = \omega_{r0}\Delta t = 0$$

$$l_1 = \frac{d\omega_r}{dt}\bigg|_0 \Delta t = \frac{\pi f}{H}(0.8 - 0.742\sin 0.427°) \times 0.05 = 0.774$$

Second estimate:

$$k_2 = l_1\Delta t = 0.774 \times 0.05 = 0.0387$$

$$l_2 = \frac{\pi f}{H}[P_m - P_{e(\delta_0 + k_1)}] \times 0.05 = \frac{\pi f}{H}(0.8 - 0.742\sin 0.427°) \times 0.05$$

$$= \frac{\pi \times 50}{5}(0.8 - 0.742\sin 0.427°) \times 0.05 = 0.774$$

$P_{e(\delta_0 + k_1)}$ is the machine power obtained from power flow using $\delta_1 = \delta_0 + k_1$.

$$\delta_1 = 0.427 + \frac{0 + 0.0387}{2} = 0.4463 \text{ rad.} = 25.571°$$

$$\omega_{r1} = \omega_{r0} + \frac{l_1 + l_2}{2} = \frac{0.774 + 0.774}{2} = 0.774 \text{ rad./s}$$

Fourth order Runge–Kutta method:

First estimate is the same as calculated in second order method.

Second estimate:

$$k_2 = \frac{l_1}{2}\Delta t = \frac{0.774}{2} \times 0.05 = 0.0194$$

$$l_2 = \frac{\pi f}{H}[P_m - P_{e(\delta_0 + k_1/2)}]\Delta t$$

$$= \frac{\pi \times 50}{5}\left[0.8 - 0.742 \sin\left(0.427 + \frac{0}{2}\right)\right] \times 0.05 = 0.774$$

Third estimate:

$$k_3 = \frac{l_2}{2}\Delta t = \frac{0.774}{2} \times 0.05 = 0.0194$$

$$l_3 = \frac{\pi f}{H}[P_m - P_{e(\delta_0 + k_2/2)}]\Delta t$$

$$= \frac{\pi \times 50}{5}\left[0.8 - 0.742 \sin\left(0.427 + \frac{0.0194}{2}\right)\right] \times 0.05 = 0.764$$

Fourth estimate:

$$k_4 = l_3\Delta t = 0.764 \times 0.05 = 0.0382$$

$$l_1 = \frac{\pi f}{H}[P_m - P_{e(\delta_0 + k_3)}] \times \Delta t$$

$$= \frac{\pi \times 50}{5}[0.8 - 0.742 \sin(0.427 + 0.0194)]\,0.05 = 0.754$$

Final estimates:

$$\delta_1 = \delta_0 + \frac{1}{6}(k_1 + 2k_2 + 2k_3 + k_4)$$

$$= 0.427 + \frac{1}{6}(0 + 2 \times 0.0194 + 2 \times 0.0194 + 0.0382)$$

$$= 0.4461 \text{ rad.} = 25.56°$$

$$\omega_{r1} = \omega_{r0} + \frac{1}{6}(l_1 + 2l_2 + 2l_3 + l_4)$$

$$= 0 + \frac{1}{6}(0.774 + 2 \times 0.774 + 2 \times 0.764 + 0.754) = 0.767 \text{ rad./s}$$

10.2   Multi-machine Stability Analysis

In a power system, there is a large number of generators. These generators are connected to the load centres through transformers and transmission lines. All these form a grid network. For the power system to operate under stable operating conditions, the generators have to run in synchronism. Some of the generators may swing away and try to go out of synchronism following disturbances and/or removal of such disturbances. It is, therefore, necessary to study the multi-machine stability.

The multi-machine stability mainly involves the following:

1. Y_{bus} formation (original, pre-fault, post-fault (during and after fault clearance) and reclosure of line, if any)
2. Power flow analysis
3. Stability analysis using modified Euler's method or Runge–Kutta method to solve the Swing equations.

Multi-machine stability is determined through the following steps:

Step 1   From the pre-fault power flow study determine voltage behind transient reactance for all generators. This establishes generator emf magnitudes, which remain constant during the study and also gives initial rotor angles. Compute initial generator power outputs, which are the prime mover inputs that also remain constant during the study.

Step 2   Shift network buses behind transient reactances and convert loads into equivalent admittances. ($S_L^* = V_L^* I_L = V_L^* V_L Y_L = V_L^2 Y_L$).

Step 3   Find Y_{bus} for various network conditions—original, pre-fault, post-fault and reclosure of line, if so. Eliminate all but the generating buses.

Step 4   For faulted mode, find generator outputs from power angle equations and solve Swing equations.

Step 5   The above steps are repeated for all network conditions.

Step 6   Examine Swing curves of all generators and check whether any generator is going out of synchronism.

The flow chart for multi-machine stability study using modified Euler's method is given in Figure 10.4.

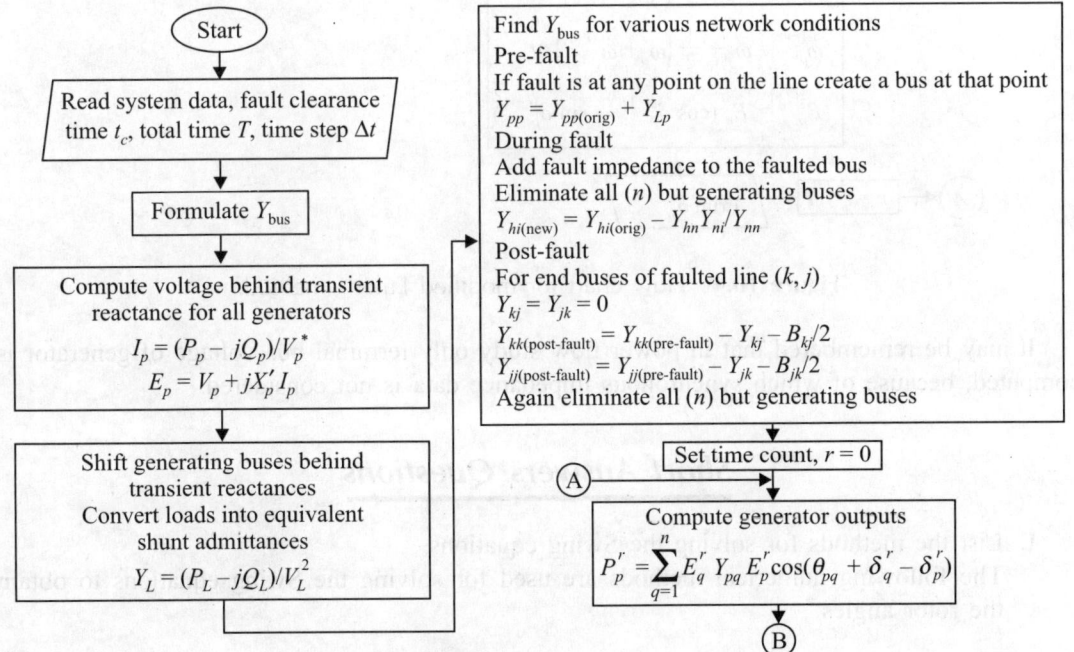

Figure 10.4 (Contd.)

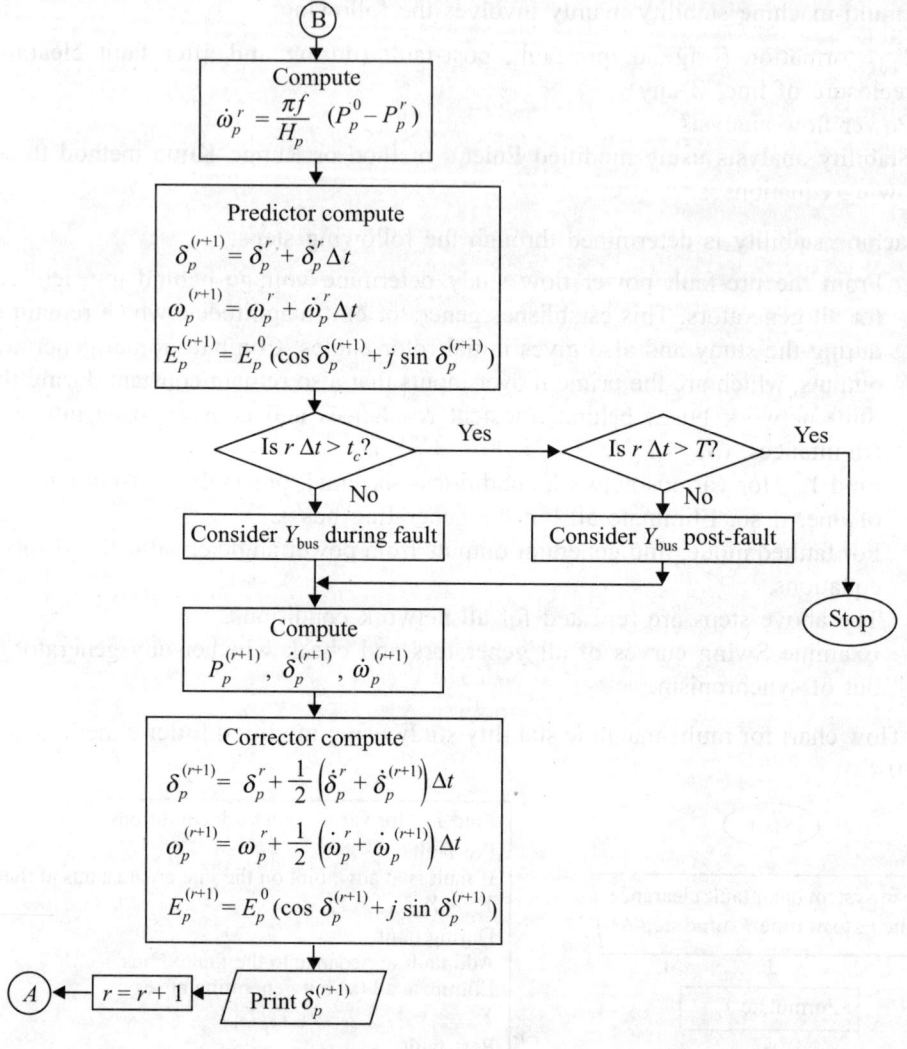

Figure 10.4 Flow chart for modified Euler's method.

It may be remembered that in power flow study only terminal bus voltage of generator is computed, because of which synchronous impedance data is not considered.

Short Answers Questions

1. List the methods for solving the Swing equations.
   The following numerical methods are used for solving the Swing equations to obtain the rotor angles:

- Euler's method
- Modified Euler's method
- Runge–Kutta method

2. What are the drawbacks of Euler's method?

   Euler's method suffers from the following shortcomings as it uses derivative at the beginning only:

   (i) The error increases as we take increasing values of t.

   (ii) If the step size Δt is large, the method may give erroneous results.

3. How does the modified Euler's method give more accurate results than Euler's method?

   The drawbacks of Euler's method are partially overcome by the use of modified Euler's method. Average of the derivatives at the beginning and the end of the step is used to find the value than the derivative at the beginning alone.

Exercise

10.1 A generator is delivering power to an infinite bus. The system data are

$$H = 5 \text{ MJ/MVA}$$
$$f = 50 \text{ Hz}$$

Power transferred before fault $= 1.8 \sin \delta = 0.8$ pu

Power transferred during fault $= 0.65 \sin \delta$

If the fault occurs at $t = 0$, find the power angle of the generator after 50 ms using both Euler's and modified Euler's methods.   [**Ans.**   26.38°, 27.75°]

GATE Questions

1. A 500 MVA, 11 kV synchronous generator has 0.2 pu synchronous reactance. The pu synchronous reactance on the base value of 100 MVA and 22 kV is:
   (a) 0.16
   (b) 0.01
   (c) 4.0
   (d) 0.25

2. A sample power system network is shown in figure. The reactances marked are in pu. The pu value of Y_{22} of the bus admittance matrix (Y_{bus}) is

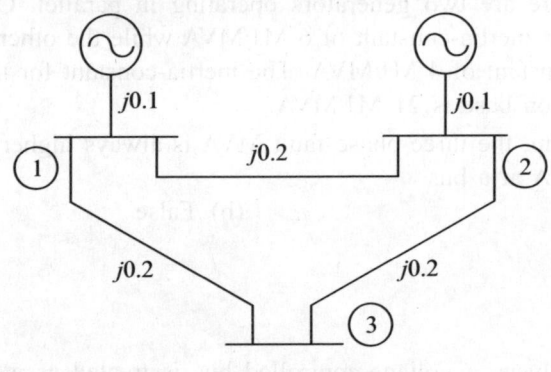

   (a) $j10.0$
   (b) $j0.4$
   (c) $-j0.1$
   (d) $-j20.0$

3. A fault occurring at the terminals of an unloaded synchronous generator operating at its rated voltage has resulted in the following values of current and voltage.

   $I_{a0} = j2.37$ pu
   $I_{a1} = j3.05$ pu
   $I_{a2} = j0.68$ pu
   $V_{a0} = V_{a1} = V_{a0} = 0.237$ pu

   the fault that has occurred is ....

[GATE 1992]

4. In load flow studies of a power system, the quantities specified at a voltage controlled bus are .... and ....

[GATE 1993]

5. The transient stability of the power system can be effectively improved by
   (a) excitation control
   (b) phase shifting transformer
   (c) single pole switching of circuit breakers
   (d) increasing the turbine valve opening
6. In load flow analysis, the load connected at a bus is represented as
   (a) constant current drawn from the bus
   (b) constant impedance connected at the bus
   (c) voltage and frequency dependent source at the bus
   (d) constant real and reactive drawn from the bus

[GATE 1994]

7. In a system, there are two generators operating in parallel. One generator, of rating 250 MVA, has an inertia-constant of 6 MJ/MVA while the other generator of 150 MVA has an inertia-constant of 4 MJ/MVA. The inertia-constant for the combined system on 100 MVA common base is 21 MJ/MVA.
8. In a power system, the three-phase fault MVA is always higher than the single-line-to-ground fault MVA at a bus
   (a) True   (b) False

[GATE 1995]

9. In load-flow analysis, a voltage-controlled bus is treated as a load bus in subsequent iteration if .... limit is violated.
10. The positive sequence component of the voltage at the point of fault in a power system is zero for a .... fault.

[GATE 1996]

11. If the reference bus is changed in two load flow runs with same system data and power obtained for reference bus taken as specified P and Q in the latter run
    (a) the system losses will be unchanged but complex bus voltage will change

   (b) the system losses will change but complex bus voltages remain unchanged

   (c) the system losses as well as complex bus voltage will change

   (d) the system losses as well as complex bus voltage will be unchanged

12. For an unbalanced fault, with paths for zero sequence currents, at the point of fault

   (a) the negative and zero sequence voltages are minimum

   (b) the negative and zero sequence voltages are maximum

   (c) the negative sequence voltage is minimum and zero sequence voltage is maximum

   (d) the negative sequence voltage is maximum and zero sequence voltage is minimum

13. During a disturbance on a synchronous machine, the rotor swings from A to B before finally settling down to a steady state at point C on the power angle curve. The speed of the machine during oscillation is synchronous at point(s)

   (a) A and B (b) A and C

   (c) B and C (d) only at C

[GATE 1997]

14. Gauss–Seidel iterative method can be used for solving a set of

   (a) linear differential equations only

   (b) inear algebraic equations only

   (c) both linear and nonlinear algebraic equations

   (d) both linear and nonlinear equations

15. For a fault at the terminal of a synchronous generator, the fault current is maximum for a

   (a) three-phase fault   (b) three-phase to ground fault

   (c) line-to-ground fault   (d) line-to-line fault

16. A 100 MVA, 11 kV, three-phase, 50 Hz, 8 pole synchronous generator has an inertia constant H equals 4 MJ/MVA. The stored energy in the rotor of the generator at synchronous speed will be $H = \dfrac{E}{G}$

   (a) 100 MJ   (b) 400 MJ

   (c) 800 MJ   (d) 12.5 MJ

[GATE 1998]

17. For the network shown in Figure, the zero sequence reactance's in pu are indicated. The zero-sequence driving point reactance of the node 3 is:

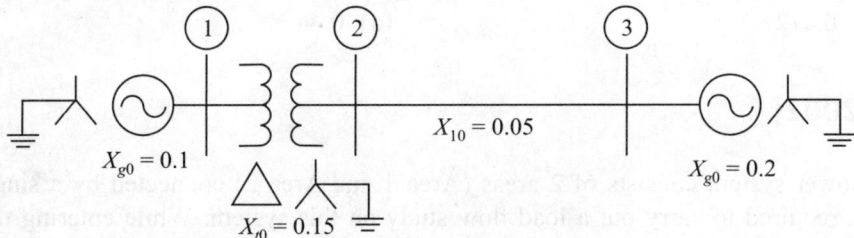

(a) 0.12   (b) 0.30

(c) 0.10   (d) 0.20

18. A power station consists of two synchronous generators A and B of ratings 250 MVA and 500 MVA with inertia 1.6 pu and 1 pu, respectively on their own base MVA ratings. The equivalent pu inertia constant for the system on 100 MVA common base is:

(a) 2.6   (b) 0.615

(c) 1.625   (d) 9.0

[GATE 1999]

19. Steady state stability of a power system is the ability of the power system to
    (a) maintain voltage at the rated voltage level
    (b) maintain frequency exactly at 50 Hz
    (c) maintain a spinning reserve margin at all times
    (d) maintain synchronism between machines and on external tie lines

[GATE 2000]

20. The severity of line-to-ground and three phase faults at the terminals of an unloaded synchronous generator is to be same. If the terminal voltage is 1.0 pu and $Z_1 = Z_2 = j0.1$ pu, $Z_0 = j0.05$ pu for the alternator, then the required inductive reactance for neutral grounding is:

(a) 0.0166 pu   (b) 0.05 pu

(c) 0.1 pu   (d) 0.15 pu

[GATE 2001]

21. A three-phase transformer has rating of 20 MVA, 220 kV (star) – 33 kV (delta) with leakage reactance of 12%. The transformer reactance (in ohms) referred to each phase of the low voltage delta-connected side is

(a) 23.5   (b) 19.6

(c) 18.5   (d) 6.5

22. 2 A 75 MVA, 10 kV synchronous generator has $X_d = 0.4$ pu. The X_d value (in pu) to a base of 100 MVA, 11 kV is

(a) 0.578   (b) 0.279

(c) 0.412   (d) 0.44

[GATE 2002]

23. A power system consists of 2 areas (Area 1 and Area 2) connected by a single tie-line. It is required to carry out a load flow study on this system. While entering the network

data, the tie-line data (connectivity and parameters) is inadvertently left out. If the load flow program is run with this incomplete data
(a) The load-flow will converge only if the slack bus is specified in Area 1
(b) The load-flow will converge only if the slack bus is specified in Area 2
(c) The load-flow will converge if the slack bus is specified in either Area 1 or Area 2
(d) The load-flow will not converge if only one slack bus is specified

24. A generator is connected to a transformer which feeds another transformer through a short feeder (see Figure 2.16). The zero sequence impedance values are expressed in pu on a common base and are indicated in Figure. The Thevenin equivalent zero sequence impedance at point B is

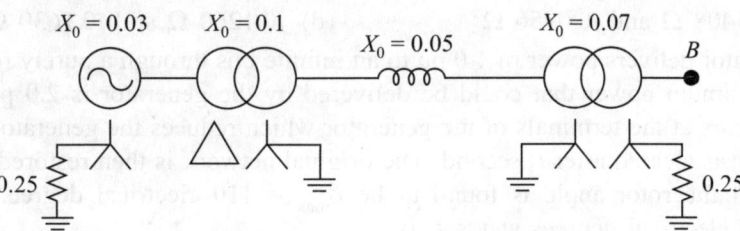

(a) $0.8 + j0.6$
(b) $0.75 + j0.22$
(c) $0.75 + j0.25$
(d) $1.5 + j0.25$

25. The graph of an electrical network has N nodes and B branches. The number of links, L, with respect to the choice of a tree, is given by
(a) $B - N + 1$
(b) $B + N$
(c) $N - B + 1$
(d) $N - 2B - 1$

26. A transmission line has a total series reactance of 0.2 pu. Reactive power compensation is applied at the midpoint of the line and it is controlled such that the midpoint voltage of the transmission line is always maintained at 0.98 pu. If voltage at both ends of the line are maintained at 1.0 pu, then the steady state power transfer limit of the transmission line is
(a) 9.8 pu
(b) 4.9 pu
(c) 19.6 pu
(d) 5 pu

[GATE 2003]

27. A power system consists of 300 buses out of which 20 buses are generator buses, 25 buses are the ones with reactive power support and 15 buses are the ones with fixed shunt capacitors. All the other buses are load buses. It is proposed to perform a load flow analysis for the system using Newton–Raphson method. The size of the Newton–Raphson Jacobian matrix is
(a) 553×553
(b) 540×540
(c) 555×555
(d) 554×554

28. The bus impedance matrix of a 4-bus power system is given by

$$Z_{bus} = \begin{bmatrix} j0.3435 & j0.2860 & j0.2723 & j0.2277 \\ j0.2860 & j0.3408 & j0.2586 & j0.2414 \\ j0.2723 & j0.2586 & j0.2791 & j0.2209 \\ j0.2277 & j0.2414 & j0.2209 & j0.2791 \end{bmatrix}$$

A branch having an impedance of $j0.2$ Ω is connected between bus 2 and the reference. Then the values of $Z_{22,new}$ and $Z_{23, new}$ of the bus impedance matrix of the modified network are respectively.
(a) $j0.5408$ Ω and $j0.4586$ Ω
(b) $j0.1260$ Ω and $j0.0956$ Ω
(c) $j0.5408$ Ω and $j0.0956$ Ω
(d) $j0.1260$ Ω and $j0.1630$ Ω

29. A generator delivers power of 1.0 pu to an infinite bus through a purely reactive network. The maximum power that could be delivered by the generator is 2.0 pu A three phase fault occurs at the terminals of the generator which reduces the generator output to zero. The fault is cleared after t_c second. The original network is then restored. The maximum swing of the rotor angle is found to be δ_{max} = 110 electrical degree. Then the rotor angle in electrical degrees at $t = t_c$ is
(a) 55
(b) 70
(c) 69.14
(d) 72.4

30. A three-phase alternator generating unbalanced voltages is connected to an unbalanced load through a three-phase transmission line as shown in Figure The neutral of the alternator and the star point of the load are solidly grounded. The phase voltages of the alternator are $E_a = 10\angle 0°$V, $E_b = 10\angle -90°$V, $E_c = 10\angle 120°$V.

The positive sequence component of the load current is

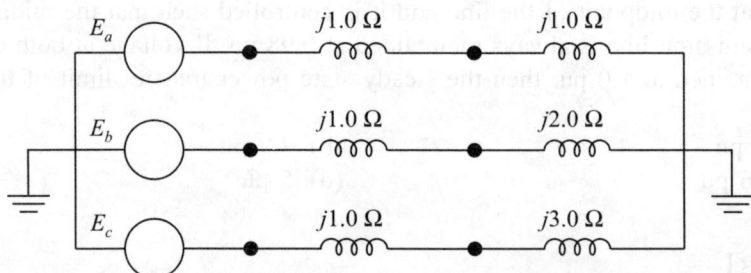

(a) $1.310\angle -107°$A
(b) $0.332\angle -120°$A
(c) $0.996\angle -120°$A
(d) $3.510\angle -81°$A

31. A 20 MVA, 6.6 kV, three-phase alternator is connected to a three-phase transmission line. The per unit positive sequence, negative sequence and zero sequence impedances of the alternator are $j0.1$, and $j0.04$ respectively. The neutral of the alternator is connected to ground through an inductive reactor of $j0.05$ pu The per unit positive, negative and zero sequence impedances of the transmission line are $j0.1$ and $j0.3$ respectively. All per unit values are based on the machine ratings. A solid ground fault occurs at one

phase of the far end of the transmission line. The voltage of the alternator neutral with respect to ground during the fault is
(a) 513.8 V   (b) 889.9 V
(c) 1112.0 V   (d) 642.2 V

32. A round rotor generator with internal voltage $E_1 = 2.0$ pu and $X = 1.1$ pu is connected to a round rotor synchronous motor with internal voltage $E_2 = 1.3$ pu and $X = 1.2$ pu. The reactance of the line connecting the generator to the motor is 0.5 pu when the generator supplies 0.5 pu power, the rotor angle difference between the machines will be
(a) 57.42°   (b) 1°
(c) 32.58°   (d) 122.58°

[GATE 2004]

33. A three-phase generator rated at 110MVA, 11 kV is connected through circuit breakers to a transformer. The generator is having direct axis sub-transient reactance $X_d'' = 26\%$ and synchronous reactance = 130%. The generator is operating at no load and rated voltage when a three-phase short circuit fault occurs between the breakers and the transformer. The magnitude of initial symmetrical rms current in the breakers will be
(a) 4.44 kA   (b) 22.20 kA
(c) 30.39 kA   (d) 38.45 kA

34. A three-phase transmission line supplies Δ-connected load Z. The conductor 'c' of the line develops an open circuit fault as shown in figure. The currents in the lines are as shown on the diagram. The positive sequence current component in line 'a' will be

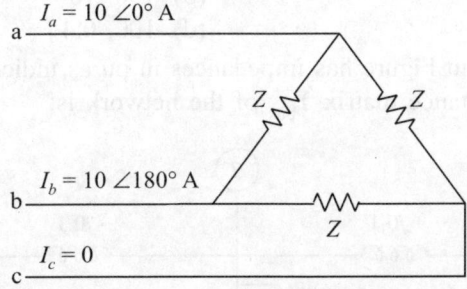

(a) 5.78∠−30°   (b) 5.78 ∠−90°
(c) 6.339∠−0°   (d) 10.00∠−30°

35. A 500 MVA, 50 Hz, three-phase turbo-generator produces power at 22 kV. Generator is Y-connected and its neutral is solidly grounded. Their sequence reactances are $X_1 = X_2 = 0.15$ and $X_0 = 0.05$ pu. It is operating at rated voltage and disconnected from the rest of the system (no load). The magnitude of the sub-transient line current for single line ground fault at the generator terminal in pu will be
(a) 2.851   (b) 3.333
(c) 6.667   (d) 8.553

36. The parameters of a transposed overhead transmission line are given as: Self reactance $X_s = 0.4$ Ω/km and Mutual reactance $X_m = 0.1$ Ω/km. The positive sequence reactance X_1 and zero sequence reactance X_0, respectively, in Ω/km are
    (a) 0.3, 0.2
    (b) 0.5, 0.2
    (c) 0.5, 0.6
    (d) 0.3, 0.6

37. A 50 Hz, 4-pole, 500 MVA, 22 kV turbo-generator is delivering rated megavoltamperes at 0.8 power factor. Suddenly a fault occurs reducing is electric power output by 40%. Neglect losses and assume constant power input to the shaft. The accelerating torque in the generator in MNm at the time of the fault will be
    (a) 1.528
    (b) 1.018
    (c) 0.848
    (d) 0.509

38. A new generator having $E_g = 1.4\angle30°$ pu [equivalent to $(1.212 + j0.70)$ pu] and synchronous reactance 'X_s' of 1.0 pu on the system base, is to be connected to a bus having voltage V_t in the existing power system. This existing power system can be represented by Thevenin's voltage $E_{th} = 0.9\angle0°$ in series with Thevenin's impedance $Z_{th} = 0.25\angle90°$ pu. The magnitude of the bus voltage V_t of the system in pu will be
    (a) 0.990
    (b) 0.973
    (c) 0.963
    (d) 0.900

[GATE 2005]

39. The pu parameters for a 500 MVA machine on its own base are: inertia $M = 20$ pu; reactance $X = 2$ pu

    The pu values of inertia and reactance on 100 MVA common base, respectively, are
    (a) 4, 0.4
    (b) 100, 10
    (c) 4, 10
    (d) 100, 0.4

40. The network shown in Figure has impedances in pu as indicated. The diagonal element Y_{22} of the bus admittance matrix Y_{bus} of the network is:

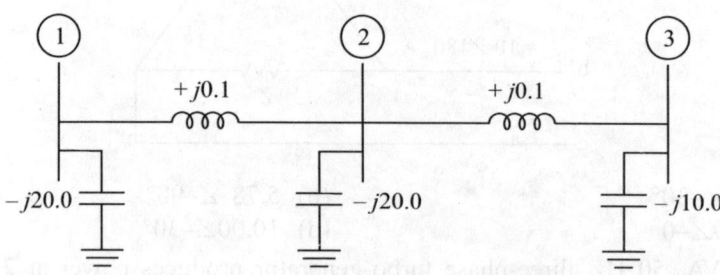

    (a) $-j19.8$
    (b) $+j20.0$
    (c) $+j0.2$
    (d) $-j19.95$

41. A generator with constant 1.0 pu terminal voltage supplies power through a step-up transformer of 0.12 pu reactance and a double-circuit line to an infinite bus bas as shown in Figure. The infinite bus voltage is maintained at 1.0 pu. Neglecting the resistances

and susceptances of the system, the steady state stability power limit of the system is 6.25 pu. If one of the double-circuit is tripped, the resulting steady state stability power limit in pu will be

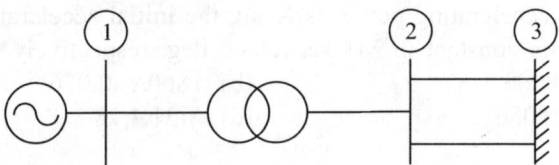

(a) 12.5 pu
(b) 3.125 pu
(c) 10.0 pu
(d) 5.0 pu

42. At a 220 kV substation of a power system, it is given that the three-phase fault level is 4000 MVA and single-line to ground fault level is 5000 MVA. Neglecting the resistance and the shunt susceptances of the system,

(i) The positive sequence driving point reactance at the bus is:
(a) 2.5 Ω
(b) 4.033 Ω
(c) 5.5 Ω
(d) 12.1 Ω

(ii) The zero sequence driving point reactance at the bus is:
(a) 2.2 Ω
(b) 4.84 Ω
(c) 18.18 Ω
(d) 22.72 Ω

[GATE 2006]

43. A generator is connected through a 20 MVA, 13.8/138 kV step down transformer, to a transmission line. At the receiving end of the line a load is supplied through a step down transformer of 10 MVA, 138/69 kV rating. A 0.72 pu load, evaluated on load side transformer ratings as base values, is supplied from the above system. For system base values of 10 MVA and 69 kV in load circuit, the value of the load (in per unit) in generator circuit will be
(a) 36
(b) 1.44
(c) 0.72
(d) 0.18

44. The Gauss–Seidel method has following disadvantages. Tick the incorrect statement.
(a) unreliable convergence
(b) slow convergence
(c) choice of slack bus affects convergence
(d) a good initial guess for voltages is essential for convergence

45. A generator feed power to an infinite bus through a double circuit transmission line. A three-phase fault occurs at the middle point of one of the lines. The infinite bus voltage is 1 pu, the transient internal voltage of the generator is 1.1 pu and the equivalent transfer admittance during fault is 0.8 pu. The 100 MVA generator has an inertia constant of 5 MJ/MVA and it was delivering 1.0 pu power prior of the fault with rotor power angle of 30°. The system frequency is 50 Hz.

(i) The initial accelerating power (in pu) will be
(a) 1.0   (b) 0.6
(c) 0.56   (d) 0.4

(ii) If the initial accelerating power is X pu, the initial acceleration in elect. deg./sec^2.,
and the inertia constant in MJ–sec./elect. deg. respectively will be
(a) 3.14X, 18   (b) 1800X, 0.056
(c) X/1800, 0.056   (d) X/31.4,18

[GATE 2007]

46. Consider a synchronous generator connected to an infinite bus by two identical parallel
transmission lines. The transient reactance X of the generator is 0.1 pu and the mechanical
power input to it is constant at 1.0 pu. Due to some previous disturbance, the rotor angle
(δ) is undergoing an undamped oscillation, with the maximum value of $\delta(t)$ equal to
130° as shown in the figure. The maximum value of the per unit line reactance X, such
that the system does not lose synchronism is

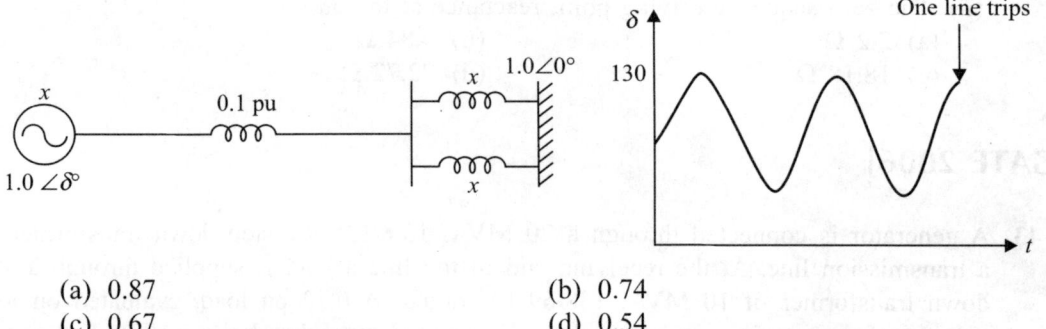

(a) 0.87   (b) 0.74
(c) 0.67   (d) 0.54

[GATE 2008]

47. A lossless single machine infinite bus power system is shown below:

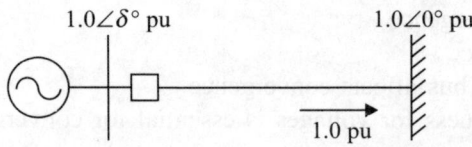

The synchronous generator transfers 1.0 pu of power to the infinite bus. Critical clearing
time of circuit breaker is 0.28 s. If another identical synchronous generator is connected
in parallel to the existing generator and each generator is scheduled to supply 0.5 pu of
power. Then the critical clearing time of the circuit breaker will
(a) reduce to 0.14 s   (b) reduce but will be more than 0.14 s
(c) remain constant at 0.28 s   (d) increase beyond 0.28 s

48. A three-phase transmission line is shown in the figure:

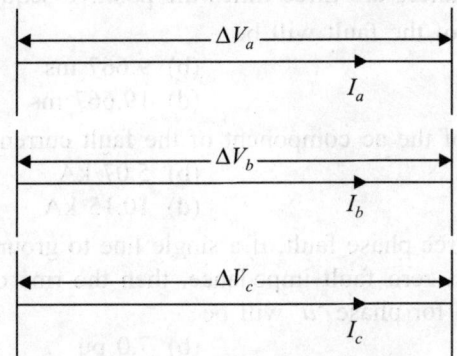

Voltage drop across the transmission line is given by the following equation:

$$\begin{bmatrix} \Delta V_a \\ \Delta V_b \\ \Delta V_c \end{bmatrix} = \begin{bmatrix} Z_s & Z_m & Z_m \\ Z_m & Z_s & Z_m \\ Z_m & Z_m & Z_s \end{bmatrix} \begin{bmatrix} I_a \\ I_b \\ I_c \end{bmatrix}$$

Shunt capacitance of the line can be neglected. If the line has the positive sequence impedance of 15 Ω and zero sequence impedance of 48 Ω, then the values of Z_s and Z_m will be

(a) $Z_s = 31.5\ \Omega$; $Z_m = 16.5\ \Omega$
(b) $Z_s = 26\ \Omega$; $Z_m = 11\ \Omega$
(c) $Z_s = 16.5\ \Omega$; $Z_m = 31.5\ \Omega$
(d) $Z_s = 11\ \Omega$; $Z_m = 26\ \Omega$

49. Consider a power system shown below

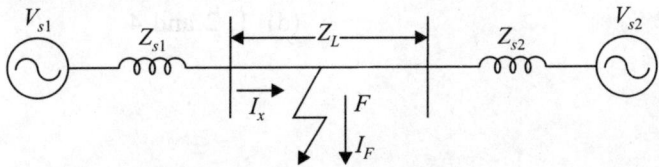

Given that:

$$V_{s1} = V_{s2} = 1 + j0 \text{ pu}$$

Positive sequence impedance are $Z_{s1} = Z_{s2} = 0.001 + j0.01$ pu and $Z_L = 0.006 + j0.06$ pu
Three-phase base MVA = 100
Voltage base = 400 kV(line-to-line)
Nominal System frequency = 50 Hz
The reference voltage for phase 'a' is defined as $v(t) = V_m \cos(\omega t)$

A symmetrical three-phase fault occurs at the centre of the line, i.e. point 'F' at time t_0. The positive sequence impedance from the source S_1 to point 'F' equals $0.004 + j0.04$. The waveform corresponding to phase 'a' fault current from bus X reveals that decaying dc offset current is negative and in magnitude at its maximum initial value. Assume that the negative sequence impedances are equal to positive sequence impedances, and the zero sequence impedances are three times the positive sequence impedance.

(i) The instant (t_0) of the fault will be
    (a) 4.682 ms   (b) 9.667 ms
    (c) 14.667 ms   (d) 19.667 ms

(ii) The rms value of the ac component of the fault current (I_x) will be
    (a) 3.59 kA   (b) 5.07 kA
    (c) 7.18 kA   (d) 10.15 kA

(iii) Instead of the three phase fault, if a single line to ground fault occurs on phase 'a' at point 'F' with zero fault impedance, then the rms of the ac component of the fault current (I_x) for phase 'a' will be
    (a) 4.97 pu   (b) 7.0 pu
    (c) 14.93 pu   (d) 29.85 pu

[GATE 2009]

50. For the Y-bus matrix of a 4-bus system given in per unit, the buses having shunt elements are

$$Y_{bus} = j \begin{bmatrix} -5 & 5 & 2.5 & 0 \\ 2 & -10 & 2.5 & 4 \\ 2.5 & 2.5 & -9 & 4 \\ 0 & 4 & 4 & -8 \end{bmatrix}$$

    (a) 3 and 4   (b) 2 and 3
    (c) 1 and 2   (d) 1, 2 and 4

[GATE 2010]

51. The zero sequence circuit of the three phase transformer shown in the figure is

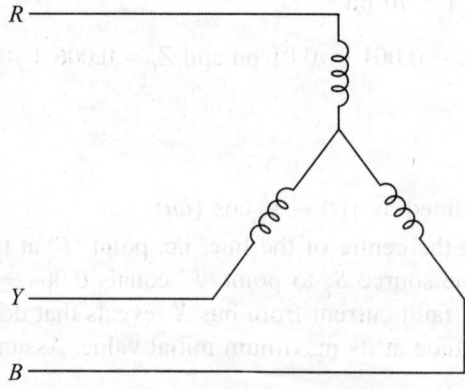
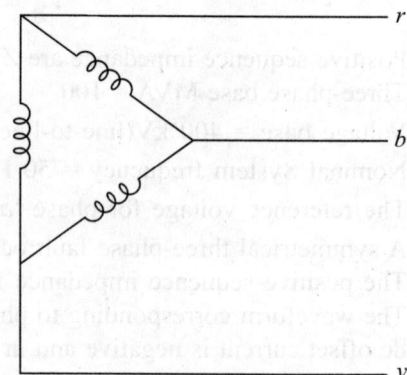

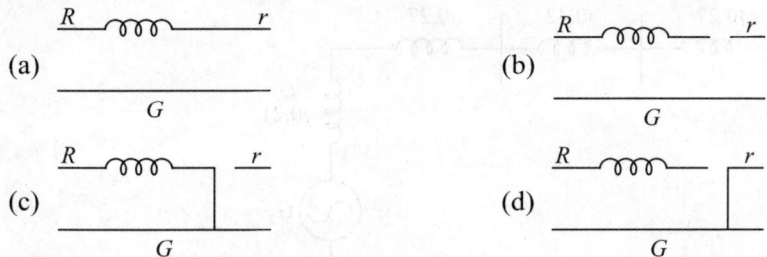

(a)

(b)

(c)

(d)

52. For the power system shown in the figure below, the specifications of the components are the following:

G_1: 25 kV, 100 MVA, $X = 9\%$

G_2: 25 kV, 100 MVA, $X = 9\%$

T_1: 25 kV/220 kV, 90 MVA, $X = 12\%$

T_1: 220 kV/25 kV, 90 MVA, $X = 12\%$

Line 1: 220 kV, 150 ohms

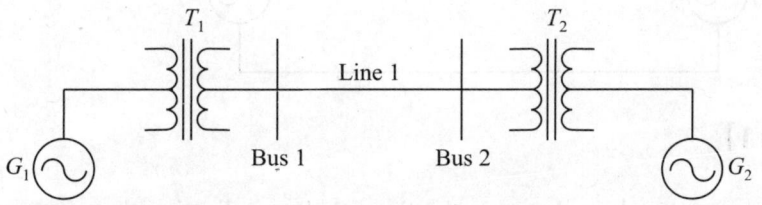

Choose 25 kV as the base voltage at the generator G_1, and 200 MVA as the MVA base. The impedance diagram is

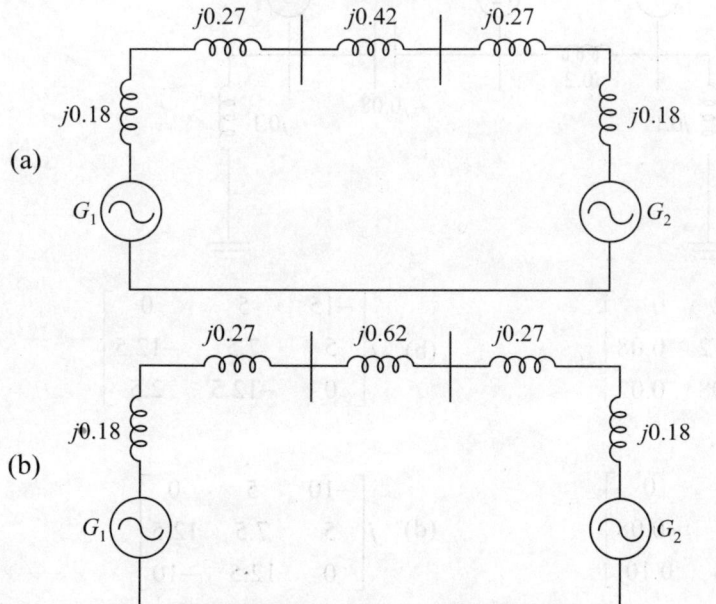

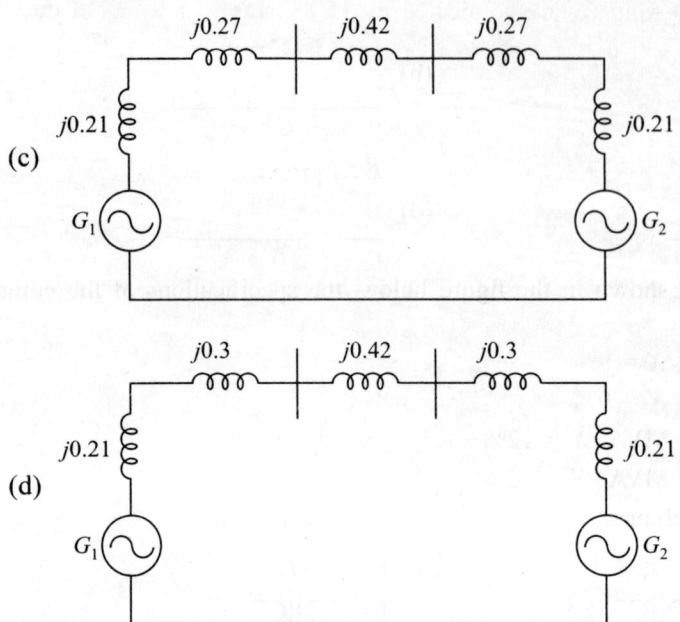

[GATE 2011]

53. A three-bus network is shown in the figure below indicating the pu impedances of each element the bus admittance matrix, Y_{bus}, of the network is

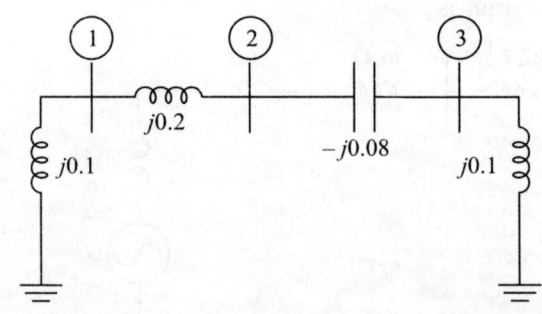

(a) $j \begin{bmatrix} 0.3 & -0.2 & 0 \\ -0.2 & 0.12 & 0.08 \\ 0 & 0.08 & 0.02 \end{bmatrix}$

(b) $j \begin{bmatrix} -15 & 5 & 0 \\ 5 & 7.5 & -12.5 \\ 0 & -12.5 & 2.5 \end{bmatrix}$

(c) $j \begin{bmatrix} 0.1 & 0.2 & 0 \\ 0.2 & 0.12 & -0.08 \\ 0 & -0.08 & 0.10 \end{bmatrix}$

(d) $j \begin{bmatrix} -10 & 5 & 0 \\ 5 & 7.5 & 12.5 \\ 0 & 12.5 & -10 \end{bmatrix}$

54. Two generator units G_1 and G_2 are connected by 15 kV line with a bus at the midpoint as shown below.

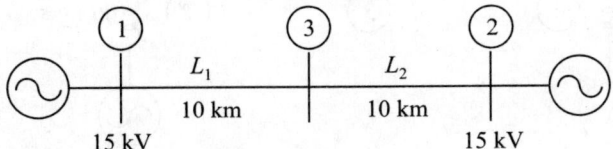

G_1 =250 MVA,15 kV, positive sequence reactance X = 25% on its own base

G_2 =100 MVA,15 kV, positive sequence reactance X = 10% on its own base

L_1 and L_2 = 10 km, positive sequence reactance = 0.225 Ω/km

   (i) For the above system, the positive sequence diagram with pu values on the 10 MVA common base is

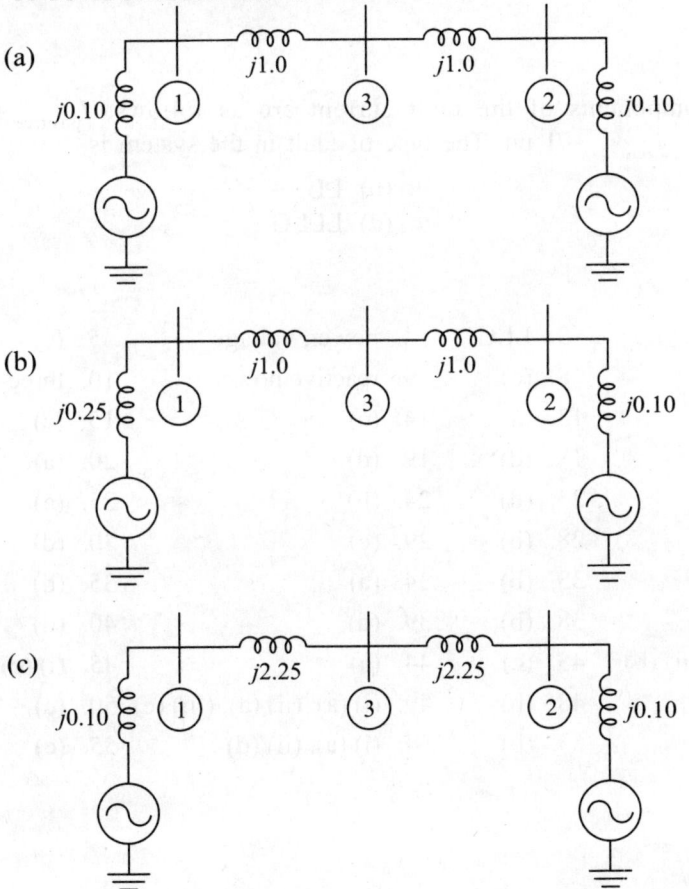

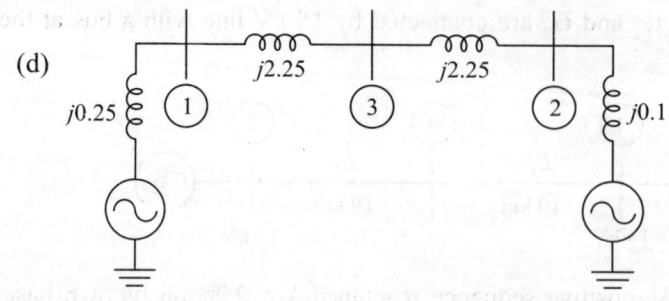

(d)

(ii) In the above system, the three phase fault MVA at bus 3 is
    (a) 82.55 MVA   (b) 85.11 MVA
    (c) 170.91 MVA   (d) 181.82 MVA

[GATE 2012]

55. The sequence components of the fault current are as follows; $I_{positive} = j1.5$ pu, $I_{negative} = -j0.5$ pu, $I_{zero} = -j1$ pu. The type of fault in the system is
    (a) LG   (b) LL
    (c) LLG   (d) LLLG

Gate Answers

1. (b)	2. (d)	3. LLG	4. power, voltage	5. (c)
6. (d)	7. 21	8. (a)	9. reactive power	10. three-phase
11. (c)	12. (a)	13. (a)	14. (c)	15. (c)
16. (b)	17. (c)	18. (d)	19. (d)	20. (a)
21. (d)	22. (d)	23. (d)	24. (b)	25. (a)
26. (a)	27. (a)	28. (b)	29. (c)	30. (d)
31. (d)	32. (c)	33. (b)	34. (a)	35. (d)
36. (d)	37. (b)	38. (b)	39. (d)	40. (d)
41. (d)	42. (i) (d), (ii) (b)	43. (c)	44. (a)	45. (i) (c), (ii) (b)
46. (c)	47. (d)	48. (b)	49. (i) (a), (ii) (a), (iii) (c)	50. (c)
51. (b)	52. (b)	53. (b)	54. (i) (a), (ii) (d)	55. (c)

References

Ahila, M. Jeraldin, *Power System Analysis*, Laxmi Publications, Chennai, 2006.

Central Electricity Authority, General Reviews, Govt. of India, New Delhi.

Das, Debapriya, *Electrical Power Systems*, New Age International (P) Ltd., Chennai, 2006.

Grainger, John J., and Stevenson, William D., Jr, *Elements of Power System Analysis*, McGraw Hill Publishing Company Ltd., New Delhi, 2003.

Gupta, B.R., *Power System Analysis and Design*, S. Chand and Company, New Delhi, 1998.

Hemalatha, K.B. and JayaChrista, S.T., *Power System Analysis*, Scitech Publications (India) Pvt. Ltd., 2006.

I. Elgerd, Olle, *Electric Energy and System Theory—An Introduction*, McGraw Hill Publishing Company Ltd., New Delhi, 1983.

Kani, A. Nagoor, *Power System Analysis*, RBA Publications, Chennai, 2004.

Karthikeyan, M., *Power System Analysis*, Dhanam Publications, Chennai.

Kothari, D.P., and Nagrath, I.J., *Modern Power System Analysis*, McGraw Hill Publishing Company Ltd., New Delhi, 2003.

Kundur, *Power System Stability and Control*, McGraw Hill Publishing Company Ltd., New Delhi, 1994.

Nagarath, I.J., and Kothari, D.P., *Power System Engineering*, McGraw Hill Publishing Company Ltd., New Delhi, 1994.

Saadat, Hadi, *Power System Analysis* McGraw Hill Publishing Company Ltd., New Delhi, 2002.

Sivanagaraju, S., and Rami Reddy, B.V., *Electrical Power System Analysis*, Laxmi Publications (P) Ltd., Chennai, 2007.

Stevenson, William D., Jr, *Power System Analysis*, McGraw Hill Publishing Company Ltd., New Delhi, 1982.

Vadhera, S.S., *Power System Analysis and Stability*, Khanna Publishers, Delhi, 2000.

Index